IoT PRODUCT DESIGN AND DEVELOPMENT

IoT PRODUCT DESIGN AND DEVELOPMENT

Best Practices for Industrial, Consumer, and Business Applications

AHMAD FATTAHI

TIBCO Software Inc.
Palo Alto, CA, USA

Registered Office
John Wiley & Sons, Inc., 111 River Street, Hoboken, NJ 07030, USA

Editorial Office
111 River Street, Hoboken, NJ 07030, USA

For details of our global editorial offices, customer services, and more information about Wiley products visit us at www.wiley.com.

Wiley also publishes its books in a variety of electronic formats and by print-on-demand. Some content that appears in standard print versions of this book may not be available in other formats.

Library of Congress Cataloging-in-Publication Data
Names: Fattahi, Ahmad, author. | John Wiley & Sons, publisher.
Title: IoT product design and development : best practices for industrial,
    consumer, and business applications / Ahmad Fattahi.
Other titles: Internet of things product design and development
    Description: Hoboken, NJ : Wiley, 2022. | Includes bibliographical
    references and index.
Identifiers: LCCN 2022017971 (print) | LCCN 2022017972 (ebook) | ISBN
    9781119787655 (cloth) | ISBN 9781119787679 (adobe pdf) | ISBN
    9781119787662 (epub)
Subjects: LCSH: Internet of things. | Product design. | New products.
Classification: LCC TK5105.8857 .F384 2022 (print) | LCC TK5105.8857
    (ebook) | DDC 004.67/8–dc23/eng/20220606
LC record available at https://lccn.loc.gov/2022017971
LC ebook record available at https://lccn.loc.gov/2022017972

Cover Design: Dana Fattahi
Cover Images: © Floaria Bicher/Getty Images, © Henrik5000/Getty Images

Set in 9/13pt STIXTwoText by Straive, Pondicherry, India

SKY10036059_091522

To My Father

Contents

Acknowledgments

I lost my father to COVID-19 during the process of writing this book. He was always the source of good and inspiration in my life. Leaving this world did not stop him from continuing to do so; his departure inspired me in a totally new way to plow ahead and hopefully help make the world a better place. Right next to him was my mother, brother, sister, and aunt who kept supporting me through the hardest days of 2021; thank you!

I've had the luxury of working with some of the best and brightest professionals in the IoT world. Richard Beeson has taught me a lot and helped me tremendously to reach where I am today. John Matranga continues to expose me to opportunities and insights that otherwise would be impossible. In fact, this book is a direct result of a joint presentation with him at Stanford University in 2019. I owe a lot to Brad Novic who is my great mentor and friend. He helped me learn about statistics and what it means to deliver value with data in a business environment. Sanjiv Patel welcomed me to Cisco and enabled me to grow in my career. His positive and growth-minded attitude will stick with me forever. I have to thank my dear friend and former colleague Jog Mahal; whenever I ran into dead ends with the legal or the bureaucratic questions at Cisco, he quickly found a way out for me with a big smile on his face. Steve Cox encouraged me a lot and played a significant role in my decision to take up the project of writing this book; thank you! On top of all of the above Richard, Brad, and Sanjiv kindly accepted the role of the reviewers for the book despite their extremely busy schedules. I also have to thank my nephew, Dana Fattahi, who eagerly helped me with his artistic prowess and designed and improved many illustrations and the cover image of this book.

I could not have completed this book if not for the tremendous amount of knowledge from a few sources. The hosts of *Stacey on IoT podcast*, Stacey Higginbotham and Kevin Tofel, kept me up to date on everything IoT on a regular basis. I also learned a lot from Dave Sluiter who teaches IoT at University of Colorado Boulder. A big shout-out goes to the whole community behind Wikipedia; I don't know what I would have done without all of the resources in this precious public source of knowledge.

The Wiley team has always been a great support. From the early days of forming the skeleton of the book to the ongoing questions about the logistics, they were always anxious to help me out. I also appreciate them accommodating me during the difficult period after my father's passing.

Last but not least, I have to appreciate my wife and best friend's support and patience throughout this gruelling process. I had to forgo many events or keep working late at night or weekends for more than a year. She kept supporting me and inspiring me with her perennial smile on her face. Thank you Mahvash!

Preface

The Internet of Things (IoT) technology is expected to generate north of $10 trillion of value globally by 2030. If you touch the consumer or business IoT ecosystem in any shape or form, there may be something of interest in this book for you. One of the major challenges of designing and implementing an Internet of Things product or service is the intrinsic multidisciplinary nature of the topic. Most products in the field have to consider significant hardware, software, data science, data engineering, and cybersecurity elements. As an example of a consumer product, a smart lock needs to be built in a compact and energy-efficient way. Some of the analytical algorithms happen on the device, while batch processes optimize larger pieces of analytics in the cloud. All such operations need to be safe and secure. The same story happens in enterprise and industrial use cases. A remote pressure sensor in the field needs to be designed to assure resilience in adverse conditions. All communication needs to happen in a secure way before IT executives give it a nod. And of course, it won't be of use if data cannot be collected at scale and fed into advanced data science algorithms.

These examples depict a complex multidisciplinary landscape that takes unique skills for a product manager, business developer, security professional, or technical product development manager. Devising the right team with the right skills and understanding the optimal dynamics among them is of crucial importance. With the expanded breadth of knowledge comes a dire need for translation of terminology amongst groups. Right expectations need to be set for all of the team members and executives in order to move in the right direction and avoid premature disappointment. Various chapters of this book guide you through all of these aspects and enable you to be more effective in your function.

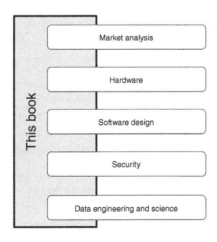

This book will set up the foundation for the successful process of going about the IoT product design and management while offering specific practical and tried recommendations before, during, and after execution. It also offers a practical recipe for the endeavor by offering technological and human-related best practices and pitfalls based on numerous real-world examples. While high-level technology, data, and business principles will be discussed, this book is not a deep technical guide that covers technological solutions in detail. Rather we cover the breadth of the technologies related to IoT at a high level. As an analogy, imagine that you are trying to design a few guided tours in two states in the United States. This book would be the equivalent of zooming out on the whole 50 states' geographical characteristics, how the federal and state regulations apply, how the seasons work in the United States, and how tours generally fail or succeed in this country. You will be able to understand who you need to connect with and understand the interests of tour guides, hikers, caterers, state park rangers, and the wildlife regulations. You may not become a restaurant owner but you would learn enough to be able to hire them and ask them the right questions. In the opposite direction, if you are a restaurant owner and like to engage with tours you can learn about their business and be more successful at negotiating a better deal. Equivalently, the reader of this book will have a much better likelihood of a successful product design and development after applying the principles in the book. You will know what aspects will be needed, what questions to ask, where to dig more deeply, and what kind of skills set you up for success. Also if you are deep into any one of the pillars of IoT products such as product development, security, hardware, software, data science, or other related fields, you will gain a valuable understanding of other skills in the ecosystem and how you can best connect with them.

WHO IS THIS BOOK FOR?

Given the multidisciplinary nature of IoT products, a wide range of skills and personas are involved. Such diversity calls for a common understanding of the big picture by the product owners and leaders of different vertical teams. Without such common understanding, the chance of siloed work, local optimization, and lack of coherence significantly increases. This book is recommended to a wide range of professionals interested in the field of IoT:

- **IoT product managers and functional leaders** who need a multidisciplinary view into the product design and development process will find this book of value. They will be able to gain insight in analyzing the market (internal or external to their organization), build the right team, and understand what questions to ask their designers and engineers.

- **Software, hardware, or data professionals** who need to gain insight into other aspects of their product while remaining focused on their area of specialty will find this book insightful. The resulting mutual understanding puts them in a great position to understand the nuances of their extended teams and enables them to better influence the direction of the business decisions.

- This book is also a great resource for **MBA students specializing in IoT or digital transformation**, as the main text or complementary book to their main textbooks.

- **Consultants** giving advice to other organizations will find this book very valuable through learning the best practices and pitfalls in an end-to-end IoT product process.

WHAT IS COVERED?

You can pick and choose what chapters you want to read and in what order. If you are already familiar with certain aspects, you can safely skip it and continue enjoying the rest of the content. Here is the chapter-by-chapter description of what we cover.

Chapter 1 provides an introduction to IoT. We see the basic concepts that make this network of connected machines special. We also discuss why the topic is so pregnant with opportunities now and over the coming years.

Chapter 2 connects IoT with the ever more important concept of digital transformation. For many industries, digital transformation is a matter of survival. For many others, it opens the door to more profitability and employee satisfaction. We make the case for IoT to be a fundamental ingredient of an effective digital transformation and how it enables IT-OT integration.

Chapter 3 reviews the market for IoT. We talk about the business opportunities and trends in the market for the technology. We will also look at several business models and partnership styles among various players in an IoT ecosystem. We consider both business and consumer markets.

Chapter 4 covers security for IoT. We cover general techniques of securing an IoT system, as well as some historical context around the topic. Some of the networking concepts that enable security are the next topic in this chapter. We also mention a few of the commonly used standards and certifications in the industry.

Chapter 5 is the longest and most comprehensive chapter in the book. It starts by introducing the process of building a team and designing an IoT product. It then moves on to a few important concepts in IoT such as sensors, file systems, machine learning, networking, and wireless communication standards. We conclude the chapter by covering two examples and a short review of managing an IoT product's lifecycle.

Chapter 6 leverages the well-known CRISP-DM process and walks you through the high-level process of building a data product. You will learn about various aspects of building a team that is set up for success and some of the most common pitfalls along the way.

Chapter 7 closes the conversation and offers some brief concluding remarks.

Chapter 1

Introduction to IoT

In this chapter, we will start to dig into the concept of Internet of Things (IoT), its importance, and the confluence of factors that have led to the adoption wave in recent years. It's important to note that certain definitions and concepts may still be the topic of debate among semanticists and engineers. The range of different views is quite wide. What matters most in this ever-changing landscape is grasping the concepts behind the terms and building a view of the range of business opportunities and learning about the technology landscape.

WHAT IS IoT?

You can argue that from the first day in the 1970s when ARPANET connected four computers together, the IoT was born. However, the common accepted definition of the concept refers to the connected network of machines (things) with limited human intervention. When most people hear of the "Internet," they imagine a human working on a gadget. However, the IoT is about connecting machines together in semi-automated ways that are uniquely different from human-dependent networks. Note that a machine does not necessarily have to be smart or capable of making decisions in order to be part of an IoT network. A pretty, simple light bulb that is connected through short-range radio signals and can follow simple instructions remotely can very well be considered part of an IoT network. Therefore, there is a distinction between connectedness and intelligence. Connectedness is sufficient for being part of the IoT network.

Usually, such constructs are capable of scaling in at least one of several dimensions: in typical personal wellness applications, wearable devices live on people's wrists across

IoT Product Design and Development: Best Practices for Industrial, Consumer, and Business Applications, First Edition. Ahmad Fattahi.
© 2023 John Wiley & Sons, Inc. Published 2023 by John Wiley & Sons, Inc.

several countries. Such devices amount to potentially millions of devices that are all "connected" according to certain protocols that are efficient and secure.

Another dimension where IoT provides scale is the amount of data being analyzed. In traditional manufacturing systems, engineers and technicians make regular and frequent readings of meters and gauges to make decisions on how to run the system efficiently. How often do you think a human can read those meters and look for anomalies? And even when they do the reading can they do a good job of analyzing data every single time? By connecting manufacturing instruments through a network you can scale the volume of data that's being collected and analyzed continuously and accurately.

Another very important aspect of IoT is what it enables over time through harvesting the value of historical observations. Typical connected devices collect data from around them (think sensors) and their own operational history (think a log file). This is a crucial distinction from traditional disconnected devices where such historical views are lost. First important use for such a ledger is to rely on trends over time and draw conclusions. A good and real example of this is in the world of manufacturing: there are pumps on the shop floor that run continuously for weeks or months at a time. Maintenance of such pumps can traditionally be done in one of two ways: you schedule such maintenance on a calendar and take them down for service on clock. One typical approach is taking the mean time between failures (MTBF). Needless to say, this practice can be massively wasteful or risky if your schedule is too conservative or too loose, respectively. The other approach is to rely on experts with "trained ears" to just be present and raise a flag when a pump doesn't sound right. To automate and scale this approach, companies have collected and saved audio signals from pumps over a long-enough period of time. This raw data set, laid over by the history of pumps' failure, can enable machine learning models to classify pump audio signals into "healthy" and "about to fail" with quite good accuracy. Without these devices being equipped with IoT sensors, such analyses would not have been possible.

Note that in the example above not only the historical behavior of a device is of value, IoT systems enable the usage of data from similar machines by combining them together. This is a very important point that leads to many budding business opportunities and new revenue streams as we will discuss later in this book.

Like almost anything else in life, with these benefits come cost and risks. First of all, equipping old and new devices with sensors and connectivity imposes cost. Business leaders need to see a compelling promise of benefit, directly to the operation of the machine or resulting opportunities, before they sign off on engineering designs that include connectivity. The other risks are security and privacy. Whenever you open a gate for good reasons, bad actors may see it as an opportunity, which may create a vulnerability vector. You also need to be extremely cautious about data movement and privacy connotations. Underestimating privacy and its close cousin, security can lead to wholesome business failures.

One last point before we move on to the other aspects of IoT is that even though the IoT itself is all about connected machines, we should not forget the importance of humans. At the end of the day what an IoT system offers has to be connected to the human world through a user interface (UI). If the UI is difficult to use, faulty, not intuitive, or otherwise hard to use, the IoT system may fail to be adopted.

WHY IS IoT IMPORTANT?

We talked about what connected machines are and why their connected system is called IoT (Figure 1.1). Now you may ask why we should care. In this segment, we take a high-level look at the need and business opportunities both for businesses and consumer markets.

Looking at the history of industries, the trend is quite clear: humans figured out tools and processes to let machines take care of certain tasks while they changed their focus to the new frontiers of innovation. In some sense, the IoT is no exception to this general trend. Take our factory reading example again. Automating the process of continuously monitoring the reading of the device is a mundane task. Computers are great at taking these tasks off of humans' plate. Another example from the consumer world is how virtually impossible it is for anyone to keep tabs on their daily activities continuously while they go about their lives. A wearable device on our wrist can be our constant watchdog counting our steps and our pulse for us.

So taking care of repeated and mundane tasks both for businesses and consumers alike is a huge reason why IoT matters. But that's not all. There are a few more important value propositions to IoT.

There are certain opportunities that are uniquely tapped into by leveraging IoT systems. Take the Internet of Medical Things (IoMT) for example. Traditionally, when a

Figure 1.1 IoT is present in many aspects of our personal and business lives.

physician asks her patient to take certain medication or physical activity or measurement, they would have to wait until the next visit. And even then the reporting is only as good as how well the patient has done his job. We also know that humans sometimes skew their own reporting to look better in front of their doctor. IoMT technologies turn this error-prone and open-loop system to a much more accurate and closed-loop system. New gadgets are now used to measure patients' blood sugar level, pulse, or drug doses taken and make it readily available to the physician. This always-on access to accurate patient data is something that is uniquely possible because of IoT.

Another example is the so-called *preemptive maintenance* in the manufacturing world. It is common to either wait for an industrial asset to fail before maintaining it or maintain them on a preset schedule. In the former approach, you are inefficient because an unplanned failure of devices is quite disruptive in most cases. It also may result in longer term damage to the equipment. In the latter approach, you may very well accrue opportunity and labor cost for no reason by maintaining a very healthy piece of equipment. Not to mention that you may also be risking an unplanned failure if your schedule is less conservative. If you make the maintenance calendar too conservative, you risk inefficiency in maintenance cost. IoT can help by constantly monitoring the equipment's condition at scale and raising an alarm if it's confident that the device needs maintenance. Such rules are not always easy though. More sophisticated equipment may need diverse instrumentation and rich and multivariate analytics on data for a reliable detection. In many cases, such upfront capital expenditure is well worth it. For example, unplanned failure of certain equipment in the gold mining industry can cost upward of a million dollars.

The other major benefit of IoT systems is what we like to call the transportability of learning. When behavioral and operational data is collected over time, models of connected devices can be built based on them. These data sets are especially useful when the underlying systems are physically too complicated to approach by first principle physical models. Once enough data is collected, the entity with access to the model can apply the learnings to any brand new device that's similar enough to the original devices. This seemingly simple concept is in reality a massive factor and enabler for many industries. When Tesla boasts about its superior edge over its rivals in autonomous technology, it is in large part because they have managed to harvest data from several thousands of machines over time. The resulting model is now applicable to any new similar machine. Compare that to other technology companies without access to this army of vehicles.

This last concept is quite deep. Because of this feature of IoT, new business models have been popping up. Back before IoT was this pervasive, a company would have to buy all the equipment they needed and maintain them. The maintenance part is significant. So they had to hire an army of technicians with enough credentials. Alternatively, they had to sign service contracts with other companies, the manufacturer, or third parties, for maintenance. Connected machines have offered a much more efficient way out: businesses can now offer service contracts from a remote distance. That, in turn, enables them to scale in ways that were never possible before. In certain cases, they actually promise a minimum percentage of uptime. That relieves their customers, who are the users of the equipment, of maintaining the devices and lets them focus their energy on their core business. The service contractor, on the other hand, scales to more customers and therefore gets access to more data; more data leads to improved models. It's a virtuous cycle.

WHY NOW?

A lot needs to happen technologically before a truly connected system becomes operational. Sensors and actuators should be available and be economically feasible. Connectivity needs to be reliable and resilient for the data to move. Data aggregators and storage systems are crucial to add value over time. Analytics on large volumes of data needs to happen while adding context from potentially other data sets. And all of the above need to happen while observing security and privacy guidelines. In the subsegments below, we will discuss each of these aspects in more detail and argue what has happened recently that made IoT a possibility.

- **Sensors and actuators:** in case of sensing and measurement, the sensor should be economically available while its size and power consumption don't hinder the product design. In certain use cases, the problem is as simple as measuring a simple physical quantity such as temperature. In many other modern use cases, what we call sensing actually takes some smartness, such as detecting a specific biometric for wellness use cases. In all of these scenarios, the industry trend has been moving in the right direction. By certain measures, the global prices of sensors on average has fallen from USD 0.66 apiece in 2010 to less than USD 0.30 in 2020. While the prices have nosedived, the capabilities of smart sensors have practically made them the first line of cognition right where the data is. In these cases, the ability to combine a decent computer board (take Raspberry Pi or Arduino for example) next to the sensor for less than US $10 adds significant smartness to the sensed data right at the edge of the network. Trained machine learning models can make sophisticated classification decisions on raw audio, video, temperature, vibration, or other types of signal with extremely low latency at low price. This level of technical capacity at this physical size and price point were not available a decade ago.

- **Connectivity:** certain actions are done close to the edge of the network where the data is generated. Smart sensors and actuators make that ever more possible. However, at some point a subset of the data needs to move from where it is generated to a cloud or somewhere in between. One reason is that edge devices have limited storage. You need to store the data for model building, regulatory requirements, and analytics. Another reason is visibility. It's very common for business operators to need (close to) real-time visibility into what is happening. Reliable and resilient connectivity is another aspect that has improved significantly over the past years. Historical wired communication kept sensors on leash. With different types of architectures and communication protocols such as Bluetooth, Zigbee, Wi-Fi, 4G, and 5G on the one hand and ever more efficient electronics on the other remote sensors are able to remain connected. The connectivity sometimes happens directly to the destination and sometimes goes through a hop or two in the middle. The latter scenario is more common when power is limited or long distance communication is too expensive at the edge, so instead a gateway device gathers the data from a number of sensors and sends it to the cloud. The mesh-enabled protocols such as Zigbee enable sensors to run on battery for a long period of time.

- **Data aggregation and storage:** storage technology has become really inexpensive to the point that it is usually not the main factor in deciding architectures. Large companies such as Google are famously storing as much data as possible *now* in the hopes of *future* monetary utilization. It means the storage cost is low enough that it doesn't negate a potential use case in the future. Computerworld magazine declared in a 2017 article that 1 MB of hard disk storage in 1976 cost $1 million; it's down to 2 cents in 2017! That's 50 million times cheaper. At the time of writing these words, you can buy a consumer-grade hard drive for less than 2 cents per 1 GB. If one sampling of data takes 8 bytes and you take one sample every one second, 1 GB can hold more than four years worth of data. It's hard to argue against storing that much valuable and high-resolution data at that cost level. Although, as we see later in the book, added data streams may add other data management complexities that need to be handled properly. Matt Komorowski has a nice collection and chart (note the log scale on the *y*-axis) in Figure 1.2.

- **Analytics:** a big part of the value from IoT applications explicitly or implicitly have roots in analyzing the ensuing data. Therefore, advances in analytical algorithms and more powerful hardware are significant enablers in recent years for IoT proliferation. Take personal wellness as an example; your smart watch can track

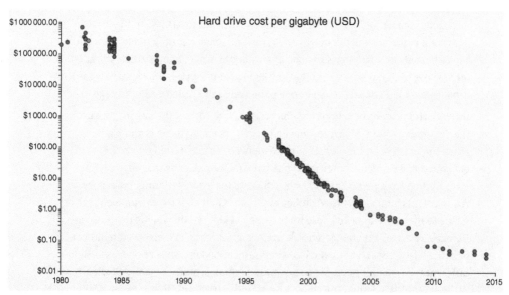

Figure 1.2 Hard drive cost per GB (*Source:* https://mkomo.com/cost-per-gigabyte).

your sleep duration and quality because in big part analytics is enabled through advanced algorithms in a small form factor that runs on very low power. Another example is when your smart home thermostat learns your patterns of behavior and adjusts the temperature accordingly to the best of your comfort. Analytics doesn't only bring exciting features to life; its role is much bigger than that. A viable IoT product needs to be *economically justified*. In many cases, such justification occurs because of a post-deployment business model that helps lower the price. For example, an emerging field in the insurance industry offers low-cost smart home gadgets and appliances to consumers in exchange for the right to acquire and use their data. Such data sets can then be sold or leased to insurance companies, manufacturers, and other third-party service companies for all sorts of business purposes. The smart home gadgets are themselves subsidized by insurers who will be able to use the data and keep their claims significantly less costly as a result. We will discuss this topic in much more detail later in this book where we argue the opportunities in data by-products.

Analytics can also be present close to the edge of the system where data is generated in the first place. Advanced machine learning algorithms can now be administered and run on inexpensive hardware to enable a multitude of use cases. Taking analytics to where the data resides is often preferred because of multiple reasons: for one, it reduces latency. The time it takes for data to make a roundtrip to a central enterprise or cloud system and back for the decision can prove crucial for certain applications such as self-driving cars. Edge analytics also reduces the security and privacy risks by shrinking the threat surface; the data travels minimally and, therefore, is less exposed to risk. Another compelling reason is that transmitting large amounts of raw data over the networks can be costly, especially if such transfer is over a cellular network. Edge computing has recently been enabled through advances in both hardware and machine learning: you can pack more sophisticated algorithms on smaller devices running on low power. Another significant use of analytics is in security. Detecting anomalies in the operation of any device or system is usually a sign of an event of interest; in many such cases, it can be due to cyberattacks or spoofed data. Low latency and efficient detection of such anomalies, enabled by analytics, make many IoT use cases secure and viable.

- **Security and privacy:** security refers to the IoT systems being immune from unauthorized manipulation both to their operation and data flow. Privacy refers to the data generated through the operation of the IoT systems being treated according to the governing policies and agreed-upon terms. These two topics are close cousins; without security you cannot expect proper privacy. Recent years have been pregnant with advances in cybersecurity, as well as security breaches and scandals. Advanced analytics, as discussed in the last subsegment, has helped systems become more secure through constant watchdog security algorithms and edge analytics. Small boards, such as a Raspberry Pi, can run classification algorithms at the edge of the system efficiently and send alarms if they detect anomalous patterns. Blockchain technology has also been creeping into more areas as a mathematically impenetrable system of records for transactions. At the same time, advanced technology leads to more connected devices, which means more opportunities for hackers. The old adage of "the chain is as strong as its weakest link" applies to this connected chain of devices. One vulnerable device can serve as the penetration point for hackers leading to widespread breaches. Security researchers have shown us several seemingly innocent connected home appliances that can act as the weakest link in your network. In one instance, they were able to show how a Wi-Fi-enabled kettle can reveal your Wi-Fi password leading to even larger beaches.

This is the right time to talk about IoT because, if done right, connected machines can actually increase our security. Also, edge analytics is now powerful enough to carry significant weight and reduce our risk. When it comes to government policies, industry consortiums, and community policing, there have recently been positive movements in unifying standards for security. We will discuss these topics in more detail when we cover security for IoT products in future chapters.

- **Climate change** has been rising in public consciousness constantly. Policy makers and public officers are being elected to help revert the dangerous trend of climate change through better policies. Municipalities, as well as federal governments, are moving in that direction to make energy production, transfer, and consumption healthier for our planet. What that means for our conversation is that more measurement needs to happen to track various environmental factors such as carbon dioxide, temperature, methane, and a whole host of other quantities. Manufacturers and the transportation industry are being required to curb and report their carbon footprint as a result of the new policies. Researchers are hungry for better ways to pinpoint the root causes of climate change with as little economical impact as possible. All of these require a stronger network of sensors that collect data and report them to central locations. That is another reason why a distributed network of inexpensive and connected sensors is an attractive proposition at this time.

Now that we have established the meaning of IoT and why it is increasingly important, we will start delving into digital transformation and the role of IoT in it.

Chapter 2

IoT and Digital Transformation

Digital transformation has been a key phrase in recent years across all corners of the economy. It refers to turning artifacts, processes, and tools into digital counterparts and starting to harvest all of the ensuing benefits. Digitally encoded content scales much better and can generate value that is orders of magnitude more than traditional ways of managing content. For example, imagine a hospital where financial transactions, doctors' briefs, lab work, prescriptions data, suppliers' data, and facilities events are all out of paper and into a digitized system. Such digitization reduces the cost of archiving and maintaining the data significantly to begin with. Without that it may be outright infeasible to record and keep all such data sets. This is only a tiny piece of the value, though. If done right, these data sets can reveal nuggets of findings that can pay dividends tenfold down the road. For example, imagine that the hospital discovers a "golden batch" where certain types of patients leave the hospital sooner with fewer side effects and return to the hospital at a lower rate compared with comparable other batches of patients. This is a question of huge importance for medical facilities. How do you go about identifying the actual causes? Is it because of the medications being used? Is there a super star physician on shift doing her miracle work? Does the room temperature play a role? Or none of the above? A digitally transformed business paves the way to make such discoveries possible. The benefits of these types of discoveries can typically dwarf the upfront costs of instrumentation. The example above was from the field of healthcare; however, the same concepts apply to virtually any business or field of industry.

IoT Product Design and Development: Best Practices for Industrial, Consumer, and Business Applications, First Edition. Ahmad Fattahi.
© 2023 John Wiley & Sons, Inc. Published 2023 by John Wiley & Sons, Inc.

There is no free lunch. Digital transformation comes at a price. If it's not done right, it can lead to lots of wasted capital. Gartner reports that close to 9 in 10 senior business leaders believe digitization is a priority for their business; however, only about 4 in 10 have been able to successfully bring it to scale. The monetary cost is only one aspect which itself can be large. The loss of faith in leadership, confused strategy, and erosion of competitive edge are among other significant costs of failed digitization efforts.

Digital transformation is much more than just turning analog documents into digits, also known as digitization. That process is part of digital transformation but certainly not all of it. It is also not only about digitizing processes even though that is part of a successful digital transformation. To do the job right, an organization needs to holistically re-envision its engagement model with its customers and employees. New opportunities and risks need to be considered. While it may make existing interactions more efficient, it may very well create brand new and fundamentally different ways that a business entity interacts with customers or potential customers.

Take the field of marketing for example. The traditional way of advertising is through open-loop engagements where customer reaction and behavior were collected en masse with lots of delay. The ad decisions for TV stations or highway billboards rely on post-advertisement interviews after previous advertisements. Those research results reach a fraction of recipients, are delayed, and are prone to errors due to the analog nature of the research. A digitally transformed way of marketing and advertising, like what online advertisers master, is fundamentally different. It tracks customer behavior at the individual level and serves the ads in a highly personalized fashion. The impact of user digital behavior, such as online searches or clicks, on the future ads is almost instantaneous. And the process is very accurate because it directly measures the users' behavior on the spot without any bias. The new process is not just a faster way of rolling out highway billboards; it's a totally new way of looking at the business of running ads using digital concepts.

What is the connection to IoT? The process of digital transformation relies heavily on machines to gather lots of data at scale, process them quickly, and make a decision quickly. In every step of this process, IoT systems are typically present and play a significant role. To continue our previous advertising example, consider the phones we carry or wearable gadgets that we wear as data collecting IoT devices on us. When machines are connected, these digital observations move to data stores where massive computation can be done at speed and scale, which wouldn't have been possible otherwise. These massive data sets over time can be the source of building models, making inferences, finding root causes, and a whole host of other applications. Decisions are then made based on the individual reading of a user's behavior. These decisions can be simple or quite complex. A user who has recently searched for eye glasses and clicked on five of them today is highly likely to be interested in, well, eye glasses. In more sophisticated scenarios, some advertisers use geographical data to make less obvious inferences. How many times have you been shocked to see an ad on your phone about a device or service about which you were talking about with a friend but never searched for it anywhere? There are many (conspiracy) theories about this practice. Here is one plausible method: during or immediately after your conversation with your friend he searches for the device or service because of your conversation. The advertiser realizes that you and the other person are very likely to have been in a conversation because you sat next to each other for a prolonged period today and left at the same time. So if one of the two searches for a product, it sends a strong signal to the advertiser that you may be interested as well.

The last leg of any decision is acting on it in a digitally (re)designed system. IoT devices play a significant role there as well. In our advertising example the ads may show up on your mobile devices. Another great example is in the world of manufacturing. Producing steel is a tricky business where multiple factors can play into the efficiency of the process. What percentage of the output turns out as scrap is a massively important factor for any steel producer. While the experience of smart technicians amassed over years goes a long way, it is by no means sufficient to uncover all opportunities. Gathered data from sensors such as temperature of the sheet, raw material qualities, thickness of the sheet, speed of the rollers, and several other factors are collected and are correlated with the quality of the outcome. The expert data scientist can then make rigorous inferences about correlations and causations leading to better products. The final corrective action in most cases also needs the machines to be connected so that the right temperature or roller speed is applied based on the readings in real time. IoT enables all aspects of this setting.

One of the most important takeaways of this chapter is that digital transformation is as much dependent on digital processes as it is on humans and culture. Without a revolution in people's culture, be it employees or customers or business leaders, the best digitization efforts are doomed to fail. That's what makes the problem trickier and more interesting at the same time. In the coming segments, we will look at the specific role that IoT and connected machines can play in a successful digital transformation, what the risks and rewards are, and why the integration of machine and human matters.

THE ROLE OF IoT IN MODERN BUSINESSES

In this section, we focus on the role of IoT toward digital transformation in modern industries, businesses, and personal lives. While everything we argued in the last section is true about the benefits of true digital transformation, or opportunities, there are risks to be considered as well. There are risks that are natural to any new business endeavor and also there are risks that are unique to IoT; we will discuss them all.

Opportunities

There are many different types of opportunities that are wholly or partially attributed to IoT. We count a number of these opportunities below while distinguishing between business versus consumer domains. We also try to recount real world stories quite a bit for further clarity.

Quality of existing service: home thermostats have existed for a long time. However, newer connected thermostats have improved the quality of service significantly. They can smartly know when you are around to keep the temperature at the comfortable level during those times. By doing so they also save energy. You can control the temperature remotely, which gives you added flexibility as a user.

Case study: in an event for data and software professionals, San Diego International Airport offered data from its facilities to improve its patrons' experience while at the airport. Some of the results were so amazing that the management of the airport directly included these seemingly simple finds in their multi-year development plans. Among these findings were data reported by temperature sensors showing that in certain areas of the terminals the temperature was lower than the comfort point of average human beings. It means, while the airport was burning extra power to keep that area cool, the result was a subpar experience for the passengers. The sensors that were exclusively driving the AC system were installed in certain areas of the terminal that had a significantly different temperature behavior. Obviously the finding was a win-win, which wouldn't have been possible without an IoT system of sensors and backend data infrastructure to historize and analyze the data.

Increased efficiency: before we can expect to fix inefficiencies, we have to be able to understand where and what those inefficiencies are. And only then we can analyze and decide what needs to be done. After that the decisions need to be acted upon. In every one of these steps, IoT can be instrumental. Distributed low-energy sensors collect data and therefore reveal inefficiencies. Compute resources close to the edge of the network or farther up in the cloud find opportunities and make decisions, and eventually IoT devices perform the decisions.

Case study: to continue our study from the San Diego International Airport above, another finding was that the hot water to certain wash rooms were maintained at the desired level for humans while they were barely used. Sensors measured water temperature, fuel consumption at the heater, and water consumption in the facility. Armed with data the data professionals were able to overlay these data sets together and prove that a significant amount of fuel was being consumed to heat the water with no clear benefit to the passengers. That led to a proposal to make the system more efficient by making the hot water system only on-demand for those far-off wash rooms. The cost of installing such an on-demand system could objectively be compared against the saved fuel cost.

Secondary by-products of data: every IoT endeavor, like other business activities, needs proper research and justification; let's call those predetermined reasons opportunity A. It is quite common that over time opportunities B and C also arise out of nowhere and thereby add to the benefits of the original project or product. Interestingly in many cases B or C dwarfs A altogether and claims the center spot in the portfolio. A classic (non-IoT) example of this phenomenon was Google+ where the Photos project was built to enable picture sharing in the social platform that Google+ offered. Over time Google Photos claimed a solid spot in the Google lineup of products long after Google+ had died.

Case study: large public venues, such as San Diego International Airport, often have occupancy sensors for facilities and security purposes. They also often provide their patrons with Wi-Fi hotspots across their large public area. Any one of these distributed networks of systems can open the door for a smarter advertising business for them (note that Wi-Fi hotspots are intrinsically not sensors; however, by analyzing the trends of connected devices, the administrators and analysts can use them as occupancy sensors because they reveal occupancy within a certain radius). Occupancy is highly correlated with eyes and ears, which is marketers' main criterion for advertising decisions. By knowing which areas are more occupied than others and how different times of the day and week compare for any spot, the owner of the venue can make smarter decisions about their advertising business on the walls or boards. They can also discover areas that otherwise would be thought of not worth the investment for advertisers. This is a truly unintended consequence of an IoT system that is originally designed for security, facilities, or Wi-Fi purposes.

Added security: let's start this item with a disclaimer! IoT implementations can act as friends or foes of the malicious actors. In this part of our story, we talk about the security benefits of properly designed IoT systems. In the subsequent segment, we take the opposite position when we discuss risks. A good analogy for this part is all the governments in the world. To keep their states safe and secure each country relies on a sophisticated network of eyes and ears to collect intelligence continuously. The less you know what is happening in your area of interest the more you are susceptible to malicious acts or even unintended incidents. The same principle applies to systems and machines. Distributed information-gathering sensors act as eyes and ears in your system and, therefore, enable the administrators to detect issues and act accordingly.

Case study: any large facilities department, such as an international airport, needs to invest significantly in its physical security. By monitoring occupancy, status of each door and window, as well as cameras that monitor people's movement, security professionals can scale themselves in ways that are not otherwise feasible due to resource constraints. A single door that is left unlocked unintentionally can lead to catastrophic security consequences. Smart connected locks can not only sense and share the status of the door but also act as actuators by remotely locking and unlocking the doors per policy.

Risks

Like most other things in life, high rewards come with risks. If IoT for digital transformation is not done right, the downsides can be significant. Add to this the fact that any such fundamental transformation is very delicate in nature; it doesn't take much to fail. Therefore any planner, executive, or stakeholder should be very wary of the risks. In the following few paragraphs, we review some of the risks facing any digital transformation program.

Poor human-machine integration: one of the fundamental risks facing IoT for digital transformation is embedded in how IoT is named: Internet of *Things*. The naming is emblematic of a common overlooked aspect of such programs where the full attention goes toward connecting machines together. In order to be justified, any program or system needs to impact humans at some point. That's why an IoT system designer has to carefully imagine, analyze, consult, and vet the integration of machines with humans and make sure it offers good user experience. The experience is often a user interface (UI) where a screen, keyboard, button, or otherwise tangible item acts as the interface with the user. Poor UI design can easily kill products or even companies in spite of their superior underlying technology. Some argue that the main propeller of Apple in consumer electronics has been its superior UI to all of its products.

An even broader concept is user experience (UX). UX refers to the whole user's journey from the moment they consider the product or service all the way to adoption and even decommissioning of the product. Successful companies spend a significant amount of resources to optimize the UX. Haven't we enjoyed products that work intuitively without forcing us to read pages of manuals or watch training videos? Isn't it great if different pieces of one system know and connect to each other without us having to configure them? On the flip side, don't we hate filling in the same information three times in a traditional doctor's office thinking "Why do I need to print my name and contact three times?"

Many teams and companies focus so much on the underlying engineering problems and the technology that they forget about this important factor. The result, even if technologically superior under the hood, may turn out to be so hard or tedious to use by the humans that it can lead to lack of adoption. Lack of adoption can quickly kill initiatives.

Over-enthusiasm for ROI: how many times have you been in a conversation where a highly energetic engineer talks about the new technology that is better than anything we have seen so far and will solve 90% of our problems at a very low price? He is very certain about his claims and can't stop advocating for it. Not using it will lead to massive opportunity loss for the business, he argues. There may not be a solid business or user study behind it yet, but he is sure that this is the technology that does it all. This is a typical scene where a new technology or product is in the news and becomes the fad. Once the number of advocates reaches a critical mass, group think kicks in and many more people start the snowballing process. There have been many examples of this hype cycle such as the dotcom bubble (which was responsible for the economic recession at the beginning of the twenty-first century), blockchain, big data, and IoT, among many more. The new major technologies create a hype cycle. The hype and over-enthusiasm lead to unwise investments, which will lead to over-skepticism. It is helpful to think of IoT through this lens and take actions that mitigate risks. A shrewd and realistic look at IoT and what it can do for the business is highly important. Figure 2.1 visualizes the usual hype cycle for any new technology.

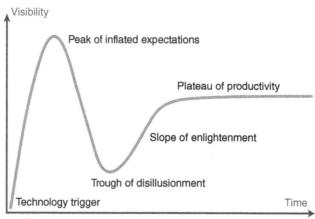

Figure 2.1 Gartner hype cycle (*Source:* https://en.wikipedia.org/wiki/Hype_cycle#/media).

The risk we are concerned about here is when the inflated expectations drive investments. By definition a subset of those inflated expectations end up being hollow. A fast and large drop in enthusiasm follows, which may in some cases result in annihilation of the technology or significant delays in adoption. In the context of IoT, we may be on to the enlightenment or real productivity phases of the cycle in certain industries. However, depending on the field of application, there may still be the risk of an enthusiasm crash. For example, in the field of industrial IoT (IIoT), there have been many proven success stories where a network of distributed sensors has led to improved visibility and optimization of the processes. In other words, IIoT led to business value. However, in many other cases, cultural inertia, technology debt, or security limitations stand in the way of transformation and lead to failed promises. One such example is transforming manual labor in factory shops where workers wear smart connected helmets with a small screen in front of one eye. The connected system senses the location and can listen to audio commands by the worker who is wearing the helmet. All the information about the equipment's history and current readings are a voice command away. These are all great and doable ideas technologically. However, rolling out this system successfully and educating the workforce to use it properly is a whole different ball game. Broad adoption of this technology will need a step change in *the way that things have been done for years,* which makes change management a challenge. Lack of adoption or, even worse, complications that may arise in the daily business routine can easily turn sweet expectations into sour experiences quickly. That's why promises and overenthusiasm have to be carefully calibrated to turn these potential failures into controlled setbacks and planned challenges. Consulting with several people in the field, starting small, and setting the expectations properly are all useful practices in this scenario. The hype cycle happens to many technologies. IoT is no exception. While in certain areas the maturity is more advanced in some other lines of business, we still need time to reach the productivity part of the journey.

Too little or too much technology: obviously, a successful IoT product or service is heavily reliant on technology. You will need enough tooling and technology for sensing physical quantities in a reliable way. Take smart doorbells as an example. While they provide a reliable and high-resolution transfer of data on a continuous basis, they require being connected to a wired and steady source of power. That thirst for power poses a limitation where they can be installed and what the customer experiences. On the other hand, wireless doorbells provide much more flexibility as to where they can be installed. The cost of that flexibility is a limitation on data transfer in addition to an occasional need for replacing batteries. Making these technological decisions properly and executing them right requires a significant amount of engineering knowledge and intuition. When it comes to more advanced derived services on top of the data, such as applying artificial intelligence to the camera feed for face recognition, there is a need for a solid backend system that is well designed to handle the computational load. Data storage for batch processing on the data is also a non-trivial technological issue. The point is that too little attention to the technological backbone of the IoT system leads to failure. This is the easier part of the problem to see.

The less clear issue arises when the technology becomes the exclusive driver. This usually happens when the founders of IoT companies or leaders of business units are themselves pure technical people. Instead of the business value analysis, the priority is given to *better technology*. While technology is unquestionably crucial, too much of it can lead to disastrous outcomes. Creating engineering artifacts for no good business reason incurs cost (that drives price point), degrades customer experience through overcomplication, delays reaching the market, exposes the product to the risk of technology decommission, and makes support and maintenance more expensive. A real-world example of this phenomenon is the Netscape web browser. While the technology was deemed very well thought-through, it took so long to develop that Microsoft captured the bulk of the market with an arguably less sophisticated product (Internet Explorer). Of course there were many other factors at play; however, time to market and other business tactics such as having the Windows leverage completely turned the competition over its head. Internet Explorer won and Netscape was acquired by AOL in 1998.

Another classic example of this phenomenon is the Netflix prize. In 2009, Netflix challenged all data scientists in the world to build a model that predicts film ratings by users based on prior observations. It was meant to "substantially improve the accuracy of predictions about how much someone is going to enjoy a movie based on their movie preferences." The winning team beat Netflix's own algorithm by a substantial margin in accuracy and won $1 million as a result. In exchange, Netflix got rights to the intellectual property. What happened to the algorithm? Nothing! The algorithm was so computationally heavy and complex that the gains would be dwarfed by the engineering cost of implementation. Again, too much technology stood in the way of implementation. Nevertheless, the groundbreaking event made machine learning popular and served as a fantastic brand-builder for Netflix.

The main lesson is while technology is extremely important to any IoT paradigm make sure that it sits in its own place. It is a mighty tool that makes a lot happen, but it is *not* the goal.

Reduced security: every new opening in a network, any new data sharing product, or any traveling packet that contains private data can become an attack surface for hackers. The product does not necessarily have to be a complex one: there are news stories where a certain type of smart coffee machine can be attacked and compromised fairly easily. The coffee machine can then expose the Wi-Fi password because it is stored as plain text. That password can let the hackers do even more damage and penetrate more devices. This is on top of all the mischief they can do on the coffee machine itself such as turning it on, showing a spooky graphic on the screen of the appliance to create fear, or otherwise create mayhem. While some of these attacks are carried out for a specific purpose such as ransom, some other ones happen only *because the hackers can do them*. The outcome for the hackers can be extorted money or bragging rights or just pure personal satisfaction. Product owners can face steep financial penalties or lose the market altogether.

Security should be part of the design process from the get go, as opposed to an afterthought. There are certain design frameworks that are extremely difficult to change after the fact. Giraffes hardly make any sound out of their mouths. One explanation is this: a long time ago their necks were shorter and they did make sounds like other mammals. The sound-making nerve from the brain to the vocal cords was wired underneath a blood vessel. Over several thousands of years they evolved into these long-necked giants that gave them the advantage of reaching fresh leaves on the top of trees. However, the nerve got stuck under the vessel; it has to travel all the way from the brain down to the bottom of the neck and then back up to the vocal cords just inches from the brain. The original flawed design was never corrected through evolution leading to a permanent shortcoming. The analogy shows how bad designs can make a product get stuck in a failed state forever.

Another important point to consider is that security applies to all aspects of an IoT product. From the moment data is being captured, it poses a certain level of risk. The more it moves around the bigger the attack vector becomes. That's why compute power close to the edge of the network is generally considered more secure than cloud computing. Also the lifecycle of the data does not end with the life of the hardware product. There are many stories in which recorded data remains vulnerable in a storage somewhere long after the device or service itself is gone. Clearly defining the processes to delete data or enable the users to permanently control this process is key. Many rules and regulations tackle these issues in the form of enforced laws (like General Data Protection Regulation or GDPR) or recommendations (like Australia's Voluntary Code of Practice). The main takeaway here is that security and privacy risks are intrinsic to IoT products. You have to acknowledge and address the risk proactively. In the following chapters, we dig into security aspects of IoT products and services in more detail.

IT'S AN INTEGRATION PROCESS: THE HUMAN ASPECTS OF DIGITAL TRANSFORMATION

IoT is officially defined as the things that are connected to each other via a network. Eventually, however, anything we build has to impact humans one way or another. Overlooking the integration of IoT solutions with humans at the right juncture can lead to suboptimal results if not total and complete failure.

Starting from the field of industrial IoT, the solutions usually end up in the field. A remote sensor that takes measurements from a gas pipe to detect leaks eventually has to offer its results to humans so they can take action. If the experience of the intended user is poorly designed, they won't get it or even campaign against such a transformation. Keep in mind that, given that this is a transformation, we are already facing an uphill battle. Humans by nature resist changes. In the most common industrial IoT examples, the intended changes go to war with well-established and decades-long practices. The battlefield is already tilted against the new processes. Not only that, many human workers see the new technology as the preamble for automation and, hence, them losing their jobs. A tiny imperfection can be blown out of proportion.

Let's look at some real stories. In one of the most established and successful industrial behemoths of the metal and mining industry, the company decided to leverage connected machines and reduce their scrap cost. The smart data professional in charge of designing the models told me several stories about how much he had to be careful about the negative feelings toward the perceived automation. Old timer process operators were used to leveraging their intuition in the metal process and make adjustments intuitively. The new models started to get on their turf by reducing the scrap ratio in an automated and cheaper way. They were not happy.

In the consumer world cybersecurity is a major concern. There is no shortage of data losses and privacy leaks in the news. Smart locks are one of the more common smart gadgets in the market. What if it gets hacked? No matter how good the device is, the human aspect of the process is crucial. A recent real story in the news revealed that a security camera was compromised due to loose security measures. It enabled a hacker, a grown male, to see and even talk to a little girl in her room at different times. The story was extremely outrageous and really bad for the manufacturer. Now the next security camera system has to fight the perception in addition to providing a quality product.

Another aspect of digital transformation in relation to humans is the actual user experience. A good product is intuitive; it just works. Bad or difficult products are hard to figure out. I have heard from several people that they opt for one smart home thermostat over the other purely because of the easier form factor or operational experience. This is completely different from comparing the features or price points of the two products. Not too many people spend the time reading lengthy user manuals. It just has to work intuitively.

In the industrial IoT realm, things are similar in nature but slightly different in form. It is expected that the solutions will be more sophisticated and may even need training. The important point is that the actual user persona of the product has to be consulted in the design phase. The people who build the product are typically software engineers, electrical engineers, or data scientists. The users of the products are typically chemical engineers, technicians, electricians, or nutrition professionals, among many others. Oftentimes, these personas don't even talk in the same language. What may be extremely obvious to a data scientist can be very vague to a marketing individual. In a real case, after months of back and forth, we realized that the intended Customer Experience users of one of our data products did not have a good grasp on the meaning of *a model*. A model, however, is the bread-and-butter concept for a data scientist. Of course, discovering this fact took a long time.

The form factor and UI/UX design is very important and yet quite nontrivial. It's way too easy to get it wrong. The US Department of Agriculture decided to set up a website for farmers and ranchers to exchange hay. Anticipating that their users may not be the most tech-savvy folks, they built the landing page with two large hyperlinks: "Have hay" and "Need hay". It can't be simpler than that, right? It appeared that half of their patrons interpreted "Have hay" as in "I have hay so I have to click here" while the other half interpreted it as "I need to connect to people who have hay, so I have to click here because I need hay."

From the users' perspective, the bottom line goal of a product is exclusively their experience with it. Nothing else should become the story. A good design should be like the air conditioner in the room. It just works. The only time it becomes the story is when it is broken!

IT-OT Integration

An important theme in the business cases is the integration or collaboration between IT and OT organizations when it comes to digital transformation. Historically, IT groups' mission has been to take care of digital assets like phones, copy machines, laptops, email servers, financial data systems, and similar services. Operational teams, depending on the industry, were in charge of the actual thing that the business was supposed to do; *line of business* some people call it: make plastic, design and build modems, or take oil out of the ground. First of all, many of these businesses are so old that the concept of data or internal networks did not even exist when the processes were created. And when the internal operational networks surfaced, through the proliferation of computers in the 1970s and 1980s, they were isolated islands that had no reason to connect to the IT systems. The OT's business was their business and none of IT's business. Not only was there no strong reason to connect them, there were security-based reasons to keep them isolated. For example, the SCADA network of a pharmaceutical company is extremely sensitive and has to be secured. Any connection to an outside network is yet another attack surface and, therefore, a security risk. A breach, inconsistency, or even loss of archived OT data can lead to batches of drugs being thrown out.

Over time, executives wanted to have real-time visibility into how different business units are doing. Smart and creative professionals found ways to cut costs or offer new services conditioned on having access to different types of data all in one place. Operational data with business context coming from finance or sales departments proved to be an exponential force multiplier by bringing business context and operational data together. Who was best prepared to make this integration happen? Usually, IT departments were tasked to do that because they are well-versed in data and cybersecurity technologies. That role in some cases was perceived as a source of power and influence for IT and against OT. OT people started to react and oppose. They saw this phenomenon as an infringement of their turf by a group of people who had no idea what their business was about. In order to make any changes to their practices, the line of business now had to seek approval from IT. IT departments, on the other hand, found OT practices unsafe and outdated. Occasional condescending personal interactions exacerbated the situation. Personal emotions developed and turf wars arose.

The point of this segment is not to offer a panacea as there is none. Executives and digital transformation designers have to start by acknowledging this challenge. Talking to different personas in advance and learning their concerns go a long way. Making people part of the transformation process, as opposed to forcing it on them, is often a powerful tool. Starting from the *why* instead of *what* or *how* is also critical. As you can see there is no magical solution that is fundamentally different from many other change management scenarios. You just need to acknowledge the challenge and proactively try to mitigate the friction.

One of the commonsense and critical steps is to create alignment between the two groups. The nature of IT and OT incentives may be very different: an OT department is rewarded by showing productivity in their line of business. When the cost of producing one unit of the product goes down or new revenue streams are added, they are rewarded. They like to jump on the newer technology as soon as it comes out or try a new methodology to invent a new product as long as they remain in control of the process. They are more likely to nod at trying new things in the interest of their rewards.

The IT department, on the other hand, is incentivized to keep the bare minimum of the business running at the minimum cost. IT departments are almost always cost centers. If they can serve the whole company at lower cost, they win. On top of that their responsibility is to keep the company secure from cyber threats. All of these factors make IT departments traditionally risk averse. Any new piece of software or new connected equipment is primarily seen as a new threat surface. Does the new system connect well with the rest of the organization's network and IT infrastructure? What is the cost of change to get all of this done?

Aligning incentives can naturally and smoothly mitigate lots of potential issues. It is the responsibility of the senior leadership of the company to define success and failure for the whole team. Even if certain parts of the organization disagree with the individual goals, they will see the rules of the game and abide by it. It would be even better if they see the thought process and reasoning. That way IT and OT departments work off of one playbook and with the same set of constraints. Success becomes everybody's success and turf wars are defused. Some of these steps may be easier said than done; however, major improvements can be achieved through these steps.

Building Trust Is Key

Transformation is as much about the culture as it is about the tools and processes. Without humans mentally shifting gears it is virtually impossible to expect a successful digital transformation. To that end organizations have to be very vigilant about the cultural aspects of their business vis à vis the transformation at hand. Individual experts need to have confidence in their leaders' vision and also trust in their colleagues in other teams. Let's discuss both of these factors.

The best strategy for transformation won't catch fire unless the rank and file believe that it's the right way to go. That's why trust and credibility of the leaders is crucial. Obviously, personal résumés of the leaders matter. On top of that the leaders should be wary of the frequency of change in direction. Change is an unavoidable part of any successful company's journey. However, any change comes at a cost. An implicit message in any change is that *the path we are on is no longer correct, so we are correcting course.* That may not be part of the official message but surely employees hear that. If the frequency of changes goes up, people start to question the prudence of their leaders. In people's minds, the leadership is throwing everything to the wall randomly to see what sticks. While digital transformation is key to survival and success of many companies, the leaders need to be very judicious about rolling out initiatives. Half-baked plans can burn enthusiasm that will be hard to get back. And when changes are to be made, the rationale needs to be explained and proper expectations need to be set in advance.

Mutual trust among peering teams is also very important. What we said in the earlier part of this segment about IT and OT departments' incentives creates a challenging starting point. Add to that the fact that oftentimes these groups speak different languages. The operations teams are experts in the products and services while the IT teams know software and networks. Having a dedicated person or group to act as level-headed, reasonable communicators between these groups is a great strategy to alleviate this problem. The last thing you want is for your IT team to talk in a condescending tone to your OT team because they don't know enough about cybersecurity, or your OT team to bash your IT team because they have no clue about what runs the business. Poised and calm communicators who know enough of both sides are very valuable in these situations.

Chapter 3

Business Models and Market Analysis

In this chapter, we are going to look at the business landscape of IoT. We take a deeper look into the opportunities and risks from the eyes of decision makers, industry professionals, and consumer experts. We also review the trends in the industry across several countries and verticals. All the trends show a fast-growing appetite to incorporate IoT as a fundamental capacity to improve business standing in several dimensions. Such dimensions include new revenue streams, optimizing processes, enhanced visibility, and enhanced security. The incentives to do so come from opportunity cost, as well as forces of business survival.

The global pandemic that started in 2019 has played an important role in expediting the above trend. Many people and businesses prefer doing business remotely, touch-free, and more smartly. IoT is in a great position to enable many such initiatives through its vast connected network of machines interacting with individuals. That is if it cannot automate the processes altogether.

BUSINESS AND INDUSTRIAL MARKET

As of 2017, it was estimated that only 1% of the industrial machines are connected to the outside world. A movement has started by a number of large industrial conglomerates to push for a smart factory in their own vertical. For example, Shell Oil developed Smart

IoT Product Design and Development: Best Practices for Industrial, Consumer, and Business Applications, First Edition. Ahmad Fattahi.
© 2023 John Wiley & Sons, Inc. Published 2023 by John Wiley & Sons, Inc.

Wells where all of the aspects of reservoir dynamics such as water flow, depth, pressure, and other operational factors are measured and monitored by connected sensors. General Electric reports that US Utility companies lose $200 billion annually due to electricity theft. True Grid by GE combined with smart meters provides the data and analytics to detect such incidents and help the utility companies to save lots of money. Rio Tinto Group is an Anglo-Australian multinational and the world's second largest metals and mining corporation. A few years ago they connected many of their drills, excavators, earth movers, and dump trucks in a way that the equipment can be operated remotely without a driver behind the wheel.

In 2019, Microsoft and the Hypothesis Group joined forces to analyze the IoT market across several industries and geographies. They produced the results in an overarching strategic report called IoT Signals report. In the following years, they repeated the process to produce the annual report. Overall, the trends show a growing and maturing environment across energy, retail, manufacturing, and healthcare verticals. The report covers the US, UK, Germany, France, China, and Japan and summarizes conversations with thousands of decision makers. The general trend shows that IoT continues to dominate a significant amount of mindshare across many industries. The improvements in artificial intelligence, edge analytics, and digital twins enable new business opportunities. At the same time, security and technological challenges persist year over year.

First measure in the Signals report is the percentage of organizations that have considered IoT in their business. This number has risen from 85% in 2019 to 91% in 2020. On the surface, it already shows a saturated market. However, the reality is far from it. The range of adoption is broken down to four stages; the landscape shows a fairly even distribution among these four stages:

- **Learn:** this is when you form study groups and send employees to find out about the business opportunities and the underlying technology. This is the least amount of commitment. The learning group can be very small, even one or two people depending on the size of the project. What is required is fast learners who have a good background on the fundamentals of the business and the technology.

- **Trial/PoC:** after finding some potential opportunities, you form a delta force to hack a quick solution together to prove the concept. This phase usually involves fast thinkers and tinkerers who can put a few different technologies together quickly. Scale and integration is not a high priority in this phase. In many cases, this phase is done with little to no fanfare to manage expectations. Getting *one result* out the door is the priority.

- **Purchase:** after a successful proof of concept, the business decides that it's time to purchase the solution. The purchase may mean buying off the shelf or alternatively paying for a custom project. This phase requires a significant buy-in from senior management because it can have large financial and process ramifications for the company. Security concerns also come into play in this phase and need to be fully addressed. This phase alone can take a long time due to its multifaceted nature. A typical vicious loop that companies find themselves in happens like this: with the PoC results in hand, the project champion asks for a budget to purchase. The decision maker asks for a confident proof of return on investment before signing off. The project champion says: "Even though I am confident in my gut about the return, I would need to build the full scale solution before I can show you the actual proof." And the decision maker says "We have higher priorities that are certain to return on our investment right now. Come back when you have the proof." These are typical catch-22 loops that happen all too often. Successful and carefully designed PoC results certainly help make inroads in the decision-making circles at this stage. Educating the executives on the merits of the technology and not assuming that they know the details is another key point. Having well-versed communicators pitch the idea (as opposed to necessarily the smartest engineer) is also a good decision. Pointing out leaps by competition in similar directions can also appeal to the competitive side of the executives. At the end of the day, like many other projects, the decision-making executive needs to take some level of risk to bet on the team and technology. This is not unlike any other new business initiative.

- **Use:** this is when the acquired technology and processes need to be adopted. Why is this important? Because integrating a new process in any business is a whole new challenge of its own. Human nature can resist any sort of change, and this is not an exception. A combination of carrots and sticks need to be in place to push the new changes forward. In the case of automating technologies, certain employees may actively oppose it because they may see it as a threat to their livelihood. Good change management with supporting executive power are often required for a successful adoption. The larger or older the organization, the more challenging the integration becomes.

In the research, 25% of the projects were in the Use phase. It means there are still lots of opportunities for the businesses to grow. Overall, between 2019 and 2021, the market shows a few percentage points worth of maturity overall. In other words, some more organizations are in the top two more advanced phases of the adoption than the previous year. Interestingly enough, about 8 in 10 organizations report that they have at least one IoT project that is already in the Use phase. This is a significant figure because often the first one or two projects are the hardest ones to pull off due to trust or knowledge gaps. We expect the flood gates to open even further and more projects to move faster down the pipe over the coming years. It clearly depicts an industry that is on the slope of enlightenment toward value delivery.

A natural question to ask is the time it takes to move from an idea to the Use stage. Across the four industries mentioned above, the median duration is about a year. This is a crucial factor for executives and project owners to consider thoroughly. Setting the expectations properly about the duration and budgeting accordingly in advance is crucial. A typical misstep is when enthusiastic technologists or executives promise over-ambitious timelines. Even though it may be done in good faith, underwhelming deliveries can kill projects.

In deciding about bringing a new product or service to the market or alternatively proposing an in-house project knowing the gravity of the market is essential. In the following segment, we go through five different verticals and discuss what constitutes the main use cases.

Manufacturing: visibility and monitoring is one of the top attractions of IoT for this sector. This is in part due to the high average age of larger manufacturing companies resulting in data and measurements being an after-thought. In many typical scenarios, the fact that a technician can see what is happening across her plant without physically inspecting every tank up close is a significant breakthrough. In more advanced cases, there are many large and small companies who are in the business of making rugged wearable smart devices so that workers on the factory floor can have up-to-date information literally in front of their eyeballs. The wearable gadgets are smart enough to pick the right assets based on the physical location of the wearer. Some of these devices, such as smart helmets, are voice enabled, so the worker can have his hands free.

Automation is a close cousin of visibility and a natural extension. Once you can see what is happening all in one place and in digital format, you can start automating manual processes effectively. This kind of automation not only reduces headcount cost but also improves quality by eradicating human error. In calculating ROI for projects, headcount and human error costs are both quite significant.

IoT can improve production planning and scheduling by bringing supply chain information on the one hand and production line status on the other all in one place. Given that modern products consist of many different pieces coming together, the product owners and shift managers can more accurately see their bottlenecks and proactively mitigate them.

Quality, compliance, and production optimization are also very attractive in this vertical. IoT value can come in at least two different forms: identifying quality issues earlier, and more specifically. For example, chip production foundries have a very complicated process to manage before the chip is tested and out for shipping. In case of a failed test, identifying where in the complex process the failure happened expedites root cause analysis and helps them increase their yield; IoT combined with artificial intelligence can exactly make that happen. By embedding connected sensors all over the supply chain and production line, a chemical engineer can see quality issues in any one of the input raw materials early and move to surgically fix the problem. This is in contrast with the old days when only a few checks were done manually over a randomized sample of products to determine issues. Meeting quality requirements and avoiding ensuing penalties are real business cases for many IoT projects.

Power and Utilities: asset maintenance in a grid is a hugely important issue. A very public case in point is the biggest utility company in the most populated state in the United States, California, called Pacific Gas and Electric (PG&E). Pacific Gas and Electric reported a $3.6 billion loss in the fourth quarter of 2019 and a $7.7 billion loss for the year as a whole, weighed down by continued costs and charges related to the wildfires that drove it into bankruptcy. What was behind this horrible episode were several fire incidents that led to loss of life and lots of property in northern California. Some of the biggest instances were traced back to faulty PG&E instruments along its grid lines. When combined with dry bushes these fires were some of the largest in state history. Since then (and after restructuring) the company has been investing in remote sensing and other mitigation measures. Many startups have emerged in this sector given the massive impact of faulty grid lines on people's livelihood and the economy. Remote and timely detection of faulty equipment for grid lines enjoys a massive market demand.

Smart metering is another hot topic of the past few years. The simple and immediate thought is to eliminate the costly and error-prone practice of sending human beings to read every single meter in a network periodically. The newer meters provide connectivity from sensors through low-power and long range or mesh communication to the Internet. The cellular unit then sends the readings back to the utility company on a regular basis. This is more accurate and less costly to run compared to the old manual process. On top of that such readings provide lots more visibility to the utility companies and the customers. The provider can see at a much more granular level when and where power consumption goes up so as to incentivize their customers to behave more economically. Shaving the peak of the power demand for any utility company translates directly to bottom line improvement; hence, they have been offering *time of use* plans to encourage shifting consumption to less demanding hours. The customers, as well, can learn a lot more about their patterns of use and save money. This new business model fundamentally relies on IoT-enabled metering. On top of that some utility companies have opened up this space to their partner ecosystem. By licensing the data usage to other third-party companies (with the customer approval), they can create an ecosystem that helps them monetize their data on top of their main line of business. Distributed energy resources (DER) also call for much more data and analytics so that renewable energy customers and the utility companies can form a distributed energy market and both benefit.

Similar to manufacturing businesses, utility companies can also benefit from IoT by automating and optimizing their production and distribution operations. The nature of many utility companies dictates operation across vast geographies that are often hard to inspect in person.

In 2019, more than a dozen utilities companies were targeted by cyberattacks. The energy sector as a whole has become a soft target for state warfare. A successful attack can disrupt economies while the attackers can remain thousands of miles away and deny any wrongdoing. IoT can be a blessing or curse when it comes to cybersecurity. A well-managed IoT system can detect anomalies fast and even take autonomous measures to mitigate the impact of the attack. On the other hand, the "rotten" code base in many remote and older systems don't get patched regularly enough, if at all. Over time, they can become a threat surface.

Retail: supply chain optimization is one of the top reasons retailers like to employ IoT. There are startup IoT companies that focus on aggregating data from thousands of different businesses together securely. One of the immediate value propositions for these businesses are highlighting bottlenecks in the supply chain. It can be a daunting task to realize why production is falling short or the cost is high without such visibility. Such IoT companies can also provide aggregated and anonymized benchmarks for each region or industry so that retailers know how they compare with other similar businesses. There is an argument to share these data sets with the supply chain companies as well. Providing such visibility helps the whole *ecosystem* to be more efficient. This idea generally gains traction when the retailer is larger and quite established with enough confidence in its processes. Opening up internal data to an external audience takes a certain level of confidence in quality and position. Smaller companies may feel threatened by the adverse effects in case something goes wrong or simply due to the competitive nature of their data. We will discuss the ecosystem effect in IoT later in this chapter.

Surveillance and security are other important reasons that retailers may like to adopt IoT. Distributed and smart cameras that detect empty shelves, shoplifting, robbery incidents, or RF tags attached to pieces of merchandise are all examples of how IoT can help retailers run more safely. Smarter technologies such as face recognition can help keep repeat offenders out. Amazon has introduced the idea of a cashier-less store where you just pay by the palm of your hand.

Inventory optimization and connected logistic chains are related topics that also help businesses. IoT-enabled trucks, gates, scales, and hand-held devices all bring the universal visibility that a business owner needs to optimize their operation.

Oil and Gas: IT security is one of the top reasons why oil and gas companies are inclined to adopt IoT. As the energy markets globally shift more toward renewable sources, oil and gas companies are under pressure to reduce costs more than ever. One of the main tools to do so is through adopting IoT. As is common in the data business, the business value of data increases exponentially when different data sources are available in one place. Additional context can make or break the value of any data set: if you look at the pure vibration on a pump, you may be able to deduce certain insights about the health and operational hours of the asset. However, if you know the same data about many other pumps, the quality and viscosity of the fluid that goes through the pump, pumps' brands and dates of last service, you can find trends and gain exponentially more value. Data connectivity comes at a price though. Well-designed IoT architectures take this data merger effect into account and turn security into an intrinsic characteristic of the information network.

Predictive maintenance is another huge factor for several manufacturing and industrial companies and in particular oil and gas organizations. Traditionally, maintenance is done in one of a few ways: wait until it fails, scheduled, or rely on human experts who can intuitively tell when an asset needs maintenance. The first two approaches are obviously wasteful and also prone to all sorts of risk. The third approach relies on subjective experience of experts, creates a non-transferable knowledge base, and doesn't scale well. For all of these reasons, artificial intelligence (AI) combined with IoT can provide artificial experts that set an alarm when an asset needs maintenance.

As public opinion about climate change and related regulations ramp up, fossil fuel companies are under an increasing level of scrutiny. Monitoring and reduction of emissions are, therefore, other important reasons why oil and gas companies can leverage IoT. Sensors can be mounted to collect, make sense of, and alert about emission quantities while meeting regulatory obligations. Access to this data would not have been possible or economical without IoT.

Healthcare: tracking inventory and staff are crucial to the healthcare industry. Given the highly regulated nature of the vertical, every asset needs to be monitored at all times.

Examples are raw material containers in a pharmaceutical company or specimen collected from patients. Similarly, workers in a clean room or nurses in a hospital need to be able to connect and talk to their colleagues quickly and seamlessly without violating physical safety regulations. IoT location trackers such as RFID tags, audio-enabled headsets that understand verbal commands, sensors that measure and report all chemical characteristics of the raw material are all invaluable assets to run a safe healthcare operation.

Given the pandemic that started in 2019, remote health has become a very hot topic in the healthcare industry. All of the aspects of providing a service remotely in a secure and safe manner are required aspects of a remote healthcare provider. One aspect is a secure communication channel between patients and providers; in the United States, the Department of Health and Human Services has extensive regulations that are usually referred to as HIPAA. The Health Insurance Portability and Accountability Act of 1996 was enacted by the 104th United States Congress and signed by President Bill Clinton in 1996. As such most healthcare providers, insurers, and other entities who handle patients' data have to protect them from breach. Monitoring patients remotely and connecting with them from the safety of their home while abiding by these regulations are all enabled by advanced IoT systems. Such systems are pervasive, fine-tuned to the task at hand, and are hardened against breach and cybersecurity attacks. For example, when a doctor needs to remotely visit their patient, the videoconferencing system needs to be encrypted end to end and be password protected. Medical sensors that monitor and store patients' blood sugar or heart rate data need to store them only locally. If they need to send them elsewhere, it has to be encrypted and also seek patient's explicit permission to do so.

Let's summarize everything that we reviewed across the main five industry verticals above (Figure 3.1). The IoT train has pretty much left the station and is humming. A strong majority of companies already have some elements of IoT in their production systems with many more to come in near future. The number of connected devices in IoT networks is anticipated to be measured in tens of billions across most forecasts. While each use case has its own nuances, success is often measured by increasing yield, reducing cost, and improving quality.

Figure 3.1 IoT's top three value propositions across various verticals.

The role of AI mixed with IoT is also a massive trend that will only accelerate over time. In some sense, IoT without AI gets reduced to measurement and visibility. While still valuable, it leaves lots of business value on the table. With the fast availability of AI at low power and low price close to the edge of the network, more AI-enabled IoT devices are expected to mushroom in every factory shop to offices and on humans' wrists. The healthcare vertical may be slightly behind this wave compared to other verticals due to the high level of regulations. The impact of the global COVID-19 pandemic has expedited this vertical to adapt itself faster and offer telehealth and remote monitoring services securely.

One of the key concepts in the IoT market is the notion of *value networks* or *business ecosystems*. We will use these two terms interchangeably. Let's start from the more familiar concept of a *value chain*. For example, a farmer grows tomatoes, sells it to a ketchup factory, who then sells the ketchup to a whole buyer such as a supermarket. In the process, there are truckers, the servicers of those trucks, spice companies who serve the ketchup factory, the bottle provider, and many more businesses. At the end of the chain is the consumer who buys the ketchup bottle. Although this is a much simplified story, the concepts apply to our conversation about IoT where the business domains encompass data sensing and acquisition, processing, transmission, storage, analysis, actuation, and sharing. That is not to mention all the afterthoughts when a totally remote industry finds yet another way to use the harvested data to offer a brand new service that is uniquely enabled through remote sensing. Arguably one of the reasons that the software market grows so much faster than the older physical industries is this ability to scale: you can much more easily replicate data that's produced and harvested once. You cannot, on the other hand, just copy another bottle of ketchup from another bottle without repeating the whole production process.

Imagine a temperature sensor that is designed to optimize the operation of a production line. There is an inner loop of value creation where a device uses the data to improve its own immediate operation. That's not all though. There are also outer loops where the larger ecosystem co-creates value: sensor builders, analytics service companies, cloud providers, and insurers can all use that data in different contexts and generate net new value. Value co-creation should happen for both the consumer and also all the players in the business network. Figure 3.2 depicts a simple visualization of the interconnected nature of the value ecosystem.

To put a framework around our discussion, let's categorize business relationships in an IoT ecosystem into the following four types:

- Cooperation: firms interact with each other and combine complementary know-how or resources to achieve a common goal.
- Competition: firms compete over similar goals that can generally translate into customer and market shares.
- Coopetition: comprises competition and cooperation happening simultaneously. For instance, mobile network operators compete over customer acquisition but cooperate on network sharing.

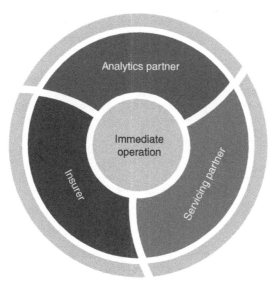

Figure 3.2 Inner and outer loops of value co-creation in IoT markets.

- Coexistence: firms exist and affect each other indirectly, without direct business interactions. They exist as part of the same industry but do not cooperate or compete over similar goals.

In our temperature sensor example, the factory shop is the center of activity. Around it there is a massive network of value co-creation. The sensor builder benefits from selling more sensors. They depend on the communication network companies, such as mobile network operators and equipment manufacturers, to transfer data to the central server or cloud. There are also the control systems (DCS) and industrial data infrastructure companies who should be compatible with the type of data to acquire it. Many companies send their data to a cloud provider eventually for storage and further analytics. That feat takes a wide area, cellular, or satellite communication. A cloud provider such as Amazon Web Services or Microsoft Azure will also be in the ecosystem. It can also be accompanied by other players in the cloud market to organize storage, streaming, or batch compute operations. Many companies tend to outsource the operation of their equipment to external service providers for better economics and scaling. Those service companies will be other significant members of the ecosystem leveraging all the data streams. Last but not least, imagine an industrial insurer or standard institution who is interested in the data once a year. The insurer optimizes their business by making sure that the operations of their insured customer is being run safely.

Does it look messy? You are not alone if you think so. The ecosystem can get quite complex with high levels of dependency and risk. This is a new concept with which the operators in the IoT market need to make peace. Business models that traditionally assume a high degree of control over the building blocks won't work well anymore. Some real examples are when an equipment builder partners with an edge AI startup

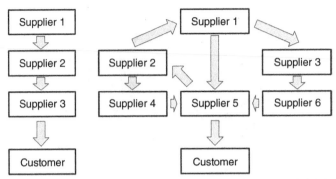

Figure 3.3 Traditional supply chain (left) vs. IoT supply network (right).

in order to offer a service. They may suddenly find themselves in hot water when the AI partner is acquired by a big company or a competitor. Given the high rate of acquisitions in the market this is a real risk. Another major factor is the dependency on certain infrastructure such as mobile communication or cloud companies. If the cloud company decides that their rates for storage or GPU will have to go up by 30% in a year the whole business model of a small IoT company may become useless. Figure 3.3 shows an ideal and simple supply chain on the left next to a more realistic view of the landscape on the right.

The point of the argument above is not to paint a bleak picture. It's rather to be realistic about the complexity of this market. Understanding which companies compete, cooperate, or coexist with each other can help. Reducing dependency on one major service provider as much as possible also mitigates the risk. If there are indications that certain big players in the market may start to expand vertically and claim a bigger portion of the whole network, other IoT business developers need to plan and adjust. When discussing business opportunities in the IoT, firms need to collaborate and be aware of novel network-centric business models. In this process, the key concern is how each firm can position itself within the business network in a way that it guarantees its profitability as part of a larger group as opposed to a single player. The challenge is that companies need to accept these new market rules where they will, individually, possess less control over the final customers and the entire value proposition. In the future, new tools to define these new ways of interaction among companies will need to be designed.

Even though there are emerging frameworks addressing business models in value networks, there are no established tools readily available in the IoT context yet. This might happen in the future and is a hot research area among business scholars. However, even in the absence of an established tool, the winning strategy is collaboration within and for the value networks considering the rich and complex market topologies. Companies need to understand how to define a business model that is aligned with the value created by the value network in the IoT ecosystem.

The Shift to Software-Based PLCS

We would be remiss if we talk about the industrial business trends of IoT without reviewing the growing trend of software-based PLCs (soft PLCs). Programmable logic controllers, or PLCs, are traditionally known as industrial black boxes. When they originally came around in the 1960s to automate industrial processes and work with sensors their underlying technology was mostly relay-based. Their communication with the outside world was mostly through proprietary I/O protocols that would require custom equipment. Needless to say, these black boxes were effective but, well, mostly black boxes. They did what they were supposed to do on day one but making any changes was virtually impossible. Also, they would typically require separate systems for the Human–Machine Interface (HMI).

In the 1980s, when the Windows operating system was invented, the hopes of making PLCs softer and less opaque increased. Companies like ASAP, Think and Do, Steeplechase Software, and Wonderware emerged in the 1990s to offer soft PLCs to the market. The value proposition was that soft PLCs are more flexible. They offer an openness to be reprogrammed much more broadly than hard PLCs of the past. They also offered an integration between the core control system and the HMI in one box. All those were appealing ideas to cut the cost and complexity. However, the operating systems of the time were not quite reliable. Frequent glitches and blue screens of death created some level of setback in the trend. That combined with the risk-averse nature of most industrial companies led to a slowdown of the adoption.

However, Linux operating systems became more reliable and the hardware capabilities got better and cheaper continually. Ethernet-based I/O made the modular design of larger systems much easier and enabled more agnostic technologies to emerge. While the capabilities improved, the cost went down. Expectations also grew based on what everyone was observing in the meteoric rise of AI and digital capabilities in the consumer market. In 2019, it was estimated that 3.5% of the industrial control market is based on soft PLCs. This share is expected to double by 2025 and reach 7%.

In such situations, a decision maker's dilemma is to stick with the tried and tested technologies of the past or to dabble with the emerging technologies. While this is a tough question to answer deterministically, we can learn important lessons from history. Kodak was the poster child of moving late to an emerging technology while having ample opportunity to do so. In 1976, the company controlled 90% of the photographic film market globally. Ironically, they were the ones who, leveraging their healthy cash flow, developed and patented the first digital camera in the world. Then the dilemma hit: on one hand they could continue bringing the new technology to the market over the coming decade or two. But the question was what that move would do to their own cash cow, the films. The fear of cannibalizing their own business caused them to not make the move. Not too long after their hesitation Asian technology companies developed less-expensive digital technology and started flooding the market with digital cameras. Long story short, in 2012 Kodak had to file for bankruptcy. The innovator's dilemma got the best of them when they could not decide to go for the new disruptive technology or stick with their reliable cash flow of the time.

Back to our soft PLC topic, the same phenomenon may be at play here. While currently the hardware-based PLC market is still strong, the trends of the market are favoring the shift to soft PLCs. The freedom to build new logic through software, containerization technology, reliable operating systems, and Ethernet-based I/O makes soft PLCs a force to be reckoned with. Market leaders in the PLC market such as ABB, Rockwell Automation, Siemens, and Schneider Electric have all been noticing the trend and investing in the new technology. Finding the sweet spot between keeping existing customers happy while inventing the future will be key to defending against disruption.

A counterargument to the market shift is that the industrial workforce is generally trained and more comfortable with the hardware-based PLCs. While there is certainly some truth there, the new and incoming workforce is much more capable of object-oriented programming and container technology than the past generation. The whole world is moving increasingly toward software and the PLC market is no exception. The IT-OT integration trend also favors soft PLCs. IT and data leaders need to be connected to and see OT data. They typically have a lot of say in the choice of OT technologies. Soft PLCs connect with the IT systems and data infrastructures more easily and flexibly.

With all of the above said, the runway for IoT technology to offer value to industries looks even more promising. By nature, IoT sensors and edge systems emerge frequently, need to be authenticated, share their data, and accept commands. Soft PLCs offer a much friendlier environment than the traditional hard PLCs because they can be programmed to work with the emerging IoT systems. If you are aiming for the trend in a few years down the road, the industrial IoT is certainly trending up in more than one way.

Example: Improve Cost and Reliability of a Power Distribution System with IoT

Imagine a power distribution company whose business is to reach tens of thousands or more small and medium size subscribers, deliver reliable electricity, measure the consumption, and bill them accordingly. There are a number of basic elements that such a business needs to do right in order to be able to continue its operations. There are also additional business opportunities for such a utility company to utilize and become more profitable. Good customer experience is another area of opportunity. In all of these areas, IoT can help.

Reliability is on top of the list of requirements. Pacific Gas and Electric is the biggest utility company in California. Established in 1905, this multibillion dollar company serves more than 5 million households in northern California. Over the years, the transmission and distribution lines aged. Transformers got older and more prone to overheating. Once in a while such faulty lines or devices started fires in remote areas that led to a fire outbreak. Led by climate change impacts, the scope of these fires grew over the 2000s and 2010s due to higher temperatures and dryer bushes. The problem got so drastic that the company was found guilty of negligence in some massive fires that claimed people's lives and billions of dollars of damage. As a result, the company declared Chapter 11 bankruptcy in January 2019. Pervasive and reliable measurement of the grid status is of paramount importance. IoT devices can reliably help. Although leveraging IoT is not a panacea, it can drastically improve the reliability of a grid operator. Fast detection of faulty lines combined with its exact location can justify a significant amount of upfront capital expenditure. Specific examples are sensors that detect and communicate smoke, higher-than normal temperature, or smartly predict a likely failure in near future.

Another aspect of operating a distribution network is being able to measure consumers' status in real time. Smart meters were introduced several years ago to provide utility companies with a granular view of usage patterns across their grid. One of the key indicators of the cost of operating a utility grid is the peak of the network. The higher the peak consumption level, the more the utility company has to spend. As a result, if consumer behavior is changed so that optional consumption is moved to off-peak hours, the utility company can pocket the profit. However, the customers should be incentivized to make such changes. The questions are:

- How do we measure the kWh consumption of each customer at scale and high granularity?
- What level of financial incentives makes sense to be offered to the customer?

IoT can help with both questions by providing the infrastructure to collect massive amounts of data at high granularity. Advanced Metering Systems (AMI) have been serving this purpose by communicating meter data to the network on a regular basis at a scale that matches the needs of urban and rural areas.

Another recent development in the distribution market is the emergence of distributed energy resources (DER). Every set of solar panels on a roof provides a new challenge or opportunity to the utility company. If not handled well, it can lead to wasted power in the grid. However, leveraging IoT, the utility company can use the historical production levels based on the weather conditions and optimize its distribution. Historical data, enabled by IoT, is key to inform the decision on when and how much to buy from consumer solar producers and how much to pay them for the practice to remain profitable.

Example: Enhance Safety, Security, and Carbon Footprint in an Airport Using IoT

An airport is like a mini city. Depending on the size and traffic of the airport the management and optimization of its operations can range from hundreds to thousands of employees. For example, Beijing International Airport employs 1600 people. That number excludes many more thousands of contractors, vendors, and airline employees.

The range of issues to be considered and solved cover a wide spectrum such as security, energy efficiency, safety, travelers' experience, vendors, monitoring airline employees, mechanical contractors, and many other aspects. We won't be opening a full-fledged conversation about an airport design here. Let's focus on energy efficiency and environmental aspects, passenger experience, security, and safety of an airport through the IoT lens.

Most airports offer water facilities in their washrooms, bathrooms, and shower facilities. Given the large size of these grids of pipes and water heaters, monitoring the condition of each faucet, pipe, or water heater becomes extremely challenging. Leakage from faucets can be detected by a network of flow detectors combined with analytics. Certain patterns in flow can be inferred as a leak, such as a faucet that is left partially open, or other types of waste. These connected IoT devices can massively reduce the cost of constant or scheduled inspections by technicians while bringing added reliability. Such leak detectors can be provisioned at scale at a few dollars each.

Another IoT use case is for rationalizing hot water consumption. Different terminals of an airport entertain different levels of passenger traffic. Keeping hot water always available at all showers or washrooms can result in a significant amount of waste in energy while adding to an airport's carbon footprint. In a study of the San Diego International Airport in 2014, it was detected that some of the less-used corners of the airport maintain hot water at all times while they were used only once or twice a day.

Airports are stressful environments for a lot of people. Intrinsically passengers are there while being sleep deprived, struck by loss of a family member, late for their flight, stranded due to flight delays, or a myriad of other potential issues. Therefore, making the environment as pleasant as possible is a priority for all airport facilities managers. One of the seemingly simple aspects is the temperature of the terminals where passengers spend hours at a time roaming around, sitting, or taking a nap. Keeping the temperature in such large facilities with sliding doors opening and closing constantly is critical and costly. In a similar study at the San Diego international Airport, sensors detected that during hot months of the year certain spots at the terminals were in fact too cold for passengers' comfort. That finding revealed a double whammy: on the one hand passenger experience was subpar and on the other hand costly energy was being wasted. Connected IoT sensors together with time-series databases and analytical tools revealed this problem and led decision makers to rationalize their facilities design.

Another use case to ease passenger experience is leveraging Wi-Fi traffic as a pseudo IoT sensor for passenger traffic. Assuming a high and consistent percentage of passengers connect to Wi-Fi hotspots at any airport, such traffic can guide facilities managers to a more accurate judgment about passenger volume per location at any time of the week. As a result, they can allocate their human and energy resources accordingly to meet passenger demands while being energy efficient.

After the 11 September 2001 attack in the United States, airports are synonymous with security. It is abundantly important to have an eye on the people entering airports and moving around. Two specific use cases can leverage IoT to enhance security. When airport, airline, and contractor employees use their badges to enter and exit secure areas, this distributed network of sensors constantly builds a timeline of people's motion inside the airport. Simple rules are always set to control permission levels. On top of that a smart aggregator, leveraging IoT data, can detect macro trends of motion at different days of the week and times of the day. As a result, any anomalous trend that deviates from the norm can be flagged as an event of interest. Another use case is distributed cameras that monitor activities at all times. A more controversial use of such cameras is facial recognition where certain faces can be programmed to raise a security incident. In these scenarios, ethical use of the technology should be applied to avoid false flags and racial discrimination.

In all of the examples above, we observed how different types of IoT systems can help make an airport a more secure, comfortable, and energy efficient place.

CONSUMER MARKET

The consumer market for IoT has been on a constant growth path for several years. Some estimates count 35 billion consumer IoT devices in the world in 2018. That number rose to 50 billion in 2020. The size of the market between the years of 2020–2025 is expected to grow at a dizzying 17% annually. While Asia offers the fastest growing geography due to market and cultural reasons, North America remains the biggest market by size.

There are several types of IoT applications today. Personal wearable device market is one giant piece that offers personal health gadgets and services through smart watches and other niche products and services. Connected cars is another massive growing segment where cars' performance, as well as passengers' experience, are targets for the IoT market. Smart home is the other big segment where entertainment, security and safety, and comfort and convenience are improved through IoT. Smart homes are typically considered the biggest chunk of this market due to several factors including the number of actions that happen in a house, offering a friendlier environment for power-hungry applications, existence of more physical space to work with compared with a human body or a car, among other reasons.

The offered categories are divided into three major groups:

1. **Hardware:** the manufacturer offers a piece of hardware such as a smartwatch or home security camera. Fitbit and Nest are good examples of such hardware manufacturers. They were both acquired by Google.

2. **Solution:** this is typically in the form of a consolidated suite of hardware and software products that collectively solve a problem or enhance someone's life. For example, a robotic vacuum cleaner that maps the layout of the house and also knows when to start vacuuming to minimize disruption inside the house is a good example. iRobot and Shark are examples in this category.

3. **Service:** in this model, a service provider offers a solution as a service. As we will see later in this section, subscription is a fast-growing business model. Examples are when a smart doorbell offers you a service to notify you when there is motion at your door and also keeps all of the video recordings in a cloud storage. Ring, which is an Amazon company, is a good example.

The state of this market is still fairly young. It means that with the myriad of opportunities and increasing consumer acceptance, there is no clear dominant winner in any of these markets. There are obviously bigger names such as Google, Amazon, Samsung, or Apple with heavy presence. However, there are still a large group of small- to medium-sized companies who are quite active. Many of the big names found their foothold in this market by acquiring many of the more successful smaller companies such as Amazon's acquisition of Ring in 2018, Samsung's acquisition of SmartThings in 2014, or Google's acquisition of Nest in 2014. Such youth in the market, as always, acts as both opportunity and challenge.

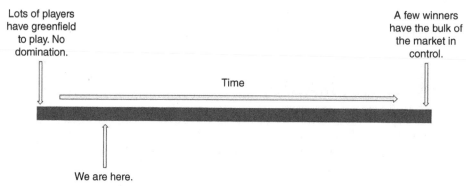

Figure 3.4 Consumer IoT market maturity level.

There is no doubt that there is a tremendous amount of green grass in this domain. If you have tried making your home a little smarter, you can easily see that certain products are still quite immature. Setups are sometimes clunky and the users need to have a certain amount of tech-savviness. Not to mention that compatibility between different products and ecosystems is also not there. That means that the price to enter this market is still relatively low. Anyone with a novel idea and decent execution can still generate some revenue.

With that said there is the flip side of the youth: chaos. Compare this scenery with the mature cell phone market. There are clear winners when it comes to the platform in iOS and Android. There are established hardware manufacturers and set expectations. Setting up a new cell phone from an old one is typically quite smooth and fast. None of that applies to the IoT market. Interoperability between devices doesn't exist. Standards are still being crafted and the governments are just about to start to pass initial security and safety regulations. Some early risers are attempting to either build their agnostic suite of products or create their proprietary ecosystems, but we are far from a steady state. Figure 3.4 puts everything in perspective in an image.

In the remainder of this segment, we will talk more about the typical sources of revenue and cost for IoT companies, elaborate some of the common types of services, discuss some of the lessons learned by startups, compare agnostic versus proprietary approaches, and finally dive into Matter, which was formerly known as Project Home over IP (CHIP).

Sources of Cost

The sources of cost for a typical consumer IoT company come from two categories. There is the typical set of costs for the operations such as office, headcount, software licenses, and the operational costs such as human resources and legal. Among these costs, typically the headcount amounts to the lion's share of the expenditure. With an ever-increasing acceptance of remote working culture remote teams can offer some advantages. One such advantage is lowering the compensation cost significantly by opening up less-expensive job markets to the employer. A typical engineer in Silicon Valley costs about three to four times as much as in India. The other advantage of choosing a distributed workforce is enlarging the pool of candidates to a much bigger pool, thereby improving the quality of the candidates. But of course the cost of running a distributed team is less efficient communication. This part applies pretty much to any business.

What can be a unique source of cost to IoT companies is the often uncertain cost of storage and compute infrastructure. Imagine a smart camera manufacturer. They have to decide how much money they are going to need to offer a certain level of service to their consumers. The service includes transferring excerpts of video captured by their cameras. These videos are often event-triggered. Right there they don't have a good way of anticipating how often their consumers will trigger events on average; is it 60 seconds per day or 600 seconds per day? Also, they have to anticipate how much storage will be needed on a regular basis to keep the footage in their cloud system. As is customary for smaller and newer companies, they like to rely on Amazon Web Services (AWS), Google Cloud Platform (GCP), or Microsoft Azure for their cloud platform. The added uncertainty is from the other side when these large behemoths decide to change their pricing strategy. In a real example, Wyze found itself in a difficult situation where the volume of video storage in AWS outgrew their worst-case anticipations. Although it was a good sign for their sales and marketing teams, it was a challenge for their engineering and platform team. They ended up negotiating from a position of more strength with AWS and inked a custom agreement with Amazon given their volume. The main takeaway here is that the uncertain nature of computation and storage needs, that are highly dependent on user behavior, is a significant source of cost for IoT companies that needs to be handled with lots of care and as much research as possible. As we will see later in this book, edge analytics enables performing more analytics closer to where data is generated. That kind of architecture, when it makes technical sense, can save bandwidth, storage, and computational cost in the cloud.

Sources of Revenue

In today's market, there are a few revenue models with a strong emphasis on one: what the customer pays at the point of sale of the gadget makes up the first revenue stream. The second potential revenue model is the recurring revenue that is often generated through upselling added services. There is also a third model that we will discuss later in this chapter: data by-products. In today's venture market, when startups are being evaluated, there is a strong emphasis on the second revenue model. Time will tell if this obsession with the subscription revenue model is truly justified, or it is a fad that will adjust. However, in certain cases the whole hardware upfront revenue is ignored by investors because in their opinion *all that matters is the recurring revenue*. The upfront cost that customers pay can help offset some of the Cost of the Goods Sold (COGS) while generally the recurring revenue is what is perceived as the growth engine.

The recurring revenue through added services can come in many different flavors. These are the types of services that offer a solution to the customer and are the hook to keep them engaged. Given the nature of many of these services, the cost of building and continually offering them keeps growing. Our video storage example is relevant where the growing volume of video footage that comes because of the service imposes significant cloud costs. You cannot keep building and selling more products to cover the increasing cost of previously sold items. That would sound more like an IoT Ponzi scheme where the business borrows against its future to make the ends meet today! Recurring subscription-based revenue can serve as a reasonable way to create sustainable revenue streams. Also, they can serve as the channel through which the manufacturer keeps its finger on the pulse of their individual users using telemetry measurements. Such measurements are extremely valuable because they provide opportunities to improve the product or services, cut features that are not of large value, create consumer loyalty through enhanced personalized offers, and reveal upsell opportunities through recommender systems. Given the criticality of these added service levels, let's spend some time explaining a number of them below.

- **Long-term data storage:** many smart home or personal wellness services include significant personal data. Examples are video footage from smart doorbells or internal security cameras. Users care about their historical data for different reasons including trend analysis, incident review for security, or health purposes. These data sets, especially when in the form of large video files, can quickly turn into a massive undertaking by the IoT manufacturer who stores these files in a cloud server. To offer the service to the users while offsetting the rising storage costs, the vendor can sell subscription services that include longer-term preservation of historical data. Oftentimes such services are bundled with other analytical and additive services for a more appealing combo as we will see below. Privacy is a very important factor to make these services acceptable by customers. Any responsible IoT manufacturer should make it easy and seamless for their users to wipe their data permanently from their servers. Imagine how much private data gets accumulated through a smart doorbell installed next to someone's door. When the user moves out, they should be able to easily wipe the device and all cloud services underneath clean of their data. Governments around the world are just catching up with rules and regulations to codify issues like this; however, there are still gaps in the regulations.

- **Contractor ads, referrals, or channeled sale of products:** why did online ads become so successful in the 2000s and thereafter? Partly because they had access to intimate behavioral data about the customers. As the old adage says "Do you know who needs a painter the most? The one who is searching for a painter." By the same token, some IoT devices provide the same or even deeper level of information about their users. For example, if I know that a customer's utility consumption has been on the rise constantly over the past two years I may be able to refer them to sustainability companies or show them an ad for certain services by the utility company. If I know that someone is regularly running during the week I can show them coupons for running shoes. Sometimes the relationship works in the reverse direction: the utility company likes to have deeper insight into their customers' behavior when it comes to heating and cooling. They sometimes subsidize certain sensors or smart thermostats and offer them at reduced cost to end users in partnership with an IoT company. The end user only gets the subsidy after their data flows to the cloud over a certain period of time. That way everybody wins: the customer becomes a smarter and more frugal energy user, the IoT vendor sells more, and the utility company gains invaluable insight about their customers and helps them shave off their peak usage. Needless to say, the usage of data has to be done responsibly and according to privacy laws and ethical principles.

- **Preventative maintenance:** this area is one of the golden standards of many IoT initiatives. Both scheduled and wait-until-broken models for maintenance are riddled with problems, inconvenience, and inefficiency. If a machine can announce when it is likely to have an issue, it can save lots of problems, time, and cost. Connected cars are one of the pioneers in this field in the consumer sector. Traditionally, every maintenance transaction for cars happens either because something failed already or that "it's been six months since the last scheduled maintenance." More modern vehicles have sensors in the more crucial spots on a car that are tied to sometimes rudimentary logic. When something is out of the normal zone, a light goes on in the car or a notification pops up in the owner's corresponding application. The technicians will then have a rich set of data stored in the car to address the issue. More IoT vendors find this whole approach appealing for their customers and a good opportunity to offer as an ongoing service.

- **Smart and improved operation:** smart sprinklers for grass watering is not a terribly new or groundbreaking idea. You can reschedule the irrigation plan from your friend's house on your phone if you see dark clouds in the sky. What can turn this into a more exciting service is to automate that logic. Imagine a number of sensors in the soil that can detect the level of moisture. The IoT vendor then combines the data with hyper-localized weather data that comes from another set of sensors and commands the sprinklers to water exactly when it is needed. This is one of the examples of advanced services where nothing is really broken. But certain mundane or wasteful tasks can now be done in a much improved way. The users should be willing to pay a subscription fee in return for more efficient irrigation and the convenience of automation. Other examples in this category include health analysis based on smart watch data. Abundant caution needs to be applied when a service is offered through a partner. Long-term commitment to customers based on third-party capabilities or relationships are prone to risk. For example, a smart camera company partnered with a young AI startup to offer face recognition to its paying customers for free. After a while, the small startup was acquired by Apple leading to a completely different cost and service model. That led to the smart camera company having to got back to the drawing board and rethink their business model due to the unsustainability of the face recognition function through Apple. In these cases, both the contract with the partner and the SLAs with end customers should be iron clad and anticipate these potential scenarios.

- **Security:** security is a major selling point for many IoT offerings in the consumer sector. There are a whole host of gadgets and sensors built specifically for security including door and window sensors, shatter sensors, indoor and outdoor security cameras, motion sensors, and many more. While these sensors offer a tremendous amount of value to the users, there are elevated levels of service that can be offered as a subscription. For example, a security camera can offer face recognition. A door open-close sensor can contact certain phone numbers or emergency services automatically. Crowd-sourced and distributed networks can issue alerts to a whole neighborhood through the applications if there is a robbery incident detected by one of the sensors or users. Since security is one of the most fundamental needs of humans this area is here to stay.

- **Third-party partnership:** besides all of the direct opportunities above, there are many chances to offer individual or aggregated data sets to third-party partners. Examples are car insurance companies or utility companies. In those examples, the utility company will offer a discount to the consumer in exchange for their data being shared through the IoT device. We will discuss some of these ideas in more detail later in the book. The one point to emphasize here is the importance of data terms and conditions. As attractive as some of these ideas may be, they can lead to disastrous public relations or legal results if done wrongly. The terms and conditions need to very clearly disclose all such use cases. On top of that the users should have easy ways to understand what they are sharing in plain language. Giving the customer the ability to opt in or out and what the default should be are other critical points that we will elaborate later.

Now that we reviewed a few different ways to generate revenue for consumer-based IoT products and services, let's shift gears to another important business decision. It is best to start with a familiar analogy. In the cellphone market, there are two dominant and successful platforms today: iOS and Android. Although each one is vastly successful, their *ecosystem designs* are vastly different. By ecosystem we mean what makes a bare bones phone attractive for consumers. In this case, all of the partners and independent developers who build millions of applications build the ecosystem for each platform. Arguably the ecosystem around the platform is as crucial, if not more, to the success of the platform than the platform itself. Consumers pay for solutions to their real-world issues and not for a brilliant operating system.

With that said Apple's approach to its ecosystem has been more of a protective one. There is one official version of the iOS that is owned and fully controlled by Apple. Any application developer or partner who wants to build an application needs to comply with the rules that are set by Apple. In contrast, Android is an open-source platform. Even though Google is the main contributor to the platform, it does not exert the same level of control over its branches. In fact there have been many branches of Android created by different companies so that they adjust the operating system with their niche requirements. We picked this example to show that there is not only one solution to this problem; in this case both approaches have passed the test of time successfully. However, each one comes with its own benefits and risks.

In the case of consumer IoT products, Wyze is a good example of trying to build their own ecosystem of products. They have been quite prolific in putting new and affordable products in the market quickly. Google has built its Nest ecosystem and Amazon has its own army of smart home products. Most of these products work very well with the other products in their own ecosystem. It is safe to assume any Nest product works seamlessly with a smart speaker and will be voice-activated. However, mixing products from different ecosystems takes some research on the part of the customer to make sure it will not lead to incompatibility. Going alone to build your own ecosystem gives you more control. You can design, from the ground up, everything in a way that is optimized for your vision. The cost of making one product compatible with everything else is minimal. There is no gap in intellectual property. You also get to increase the cost of switching for your customers, which some can see as extra stickiness for the business. The cost of doing that, though, is that you will shut yourself out of everything else in the world, or at least make integration harder to happen. Other products that are not yours may or may not be compatible with yours leading to lost customers. You would need to be big enough or powerful enough to get to a critical mass of customers and partners so that your ecosystem can survive on its own. Also, you take a bigger responsibility in keeping the quality of the products always good enough so that there is no massive exodus by customers and developers.

SmartThings, in contrast, took a niche aspect of IoT and then built them in a way that works with many other products. They tried to position themselves vertically relative to other ecosystems. Their array of special sensors connects with different systems or gets embedded in other bigger products to solve very specific problems. By doing so they opened themselves up to many more partnership opportunities. However, the cost is loss of control. If a partner wants a certain sensor to be built differently and the partner is large enough, you will have to respond. On the other hand, the customers may see your brand as less predatory by not trapping them into your ecosystem. In contrast, it leads to less organic gravity toward your ecosystem. Building more agnostic and open products is harder. You will have to learn more about existing standards and protocols continually, which takes time and resources.

A related technology that tries to mitigate incompatibility of different devices for consumers is the concept of Smart Home Hubs. In reality, many smart home devices use proprietary protocols and cannot communicate with each other. Therefore, certain companies like Samsung SmartThings offer a hub for your smart home. A hub is basically a computer that is also equipped with a number of different radio capabilities and platforms. As a result, the hub can connect with a bunch of different smart home IoT devices individually and, on the other hand, offer a unified front for the consumer to interact with them; see Figure 3.5. For example, Samsung SmartThings Hub works with sensors and devices that use low-power mesh protocols such as Zigbee and Z-Wave while also being compatible with IP-connected devices. In the other direction, it is also

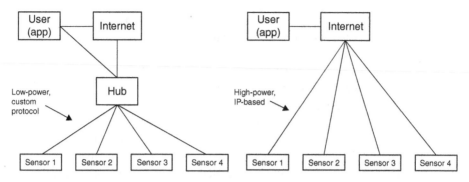

hub-based products (left) vs. hubless designs (right) in the market

Figure 3.5 Hub-based vs. hubless.

compatible with user interface platforms such as Google Assistant and Amazon Alexa for voice control and other integrations.

Another differentiator in the market is that hub-based systems allow for smaller- and lower-power sensors to become part of a smart home ecosystem. That is because hubs allow for lower-power protocols such as Zigbee to be acceptable. Internet-based hubless systems rely on each sensor to connect itself to the Internet and then from there get connected to the rest of the ecosystem. A ramification of this decision is that the sensors will typically have to rely on high-power Wi-Fi and burn through batteries quickly, or rely on Bluetooth and suffer from range limitations.

With that said we have to add that other high-level devices such as smart speakers and assistants have been adding increasingly more hub-like features to their arsenal. For a typical non-technical user, Google Assistant and Amazon Alexa can already connect to many smart devices, offer routines, and set certain scenes. Hubs used to be a necessity to have one cohesive smart home experience for many consumers; however, their position has been shifting more toward a more niche gadget for more tech-savvy users. There are still certain high-power capabilities and customizations that only a hub can offer. When compared with smart speakers, hubs are lower level devices that offer more control and more responsibility: the user sometimes needs to troubleshoot bugs, search the Internet for fellow developers' recommendations, write short snippets of code to create custom routines, or simply tolerate a little bit of a roughness when it comes to functionality.

Matter, Formerly Known as Project Connected Home Over IP (CHIP)

A natural continuation of the above topic leads us to Matter, which was originally initiated as Project Connected Home over IP or CHIP for short. After the first wave of smart home IoT devices hit the market, they brought some chaos with them. Different brands didn't work with each other, security and privacy specifications were absent or too convoluted for nonexperts to understand, and the lifecycle commitments were unclear. As a result,

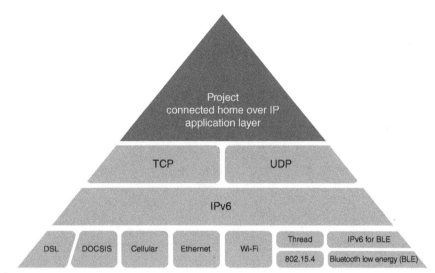

Figure 3.6 Matter's functional architecture. (*Source:* Image courtesy of Matter [formerly Connected Home over IP]).

a number of major players in the smart places market got together to define one set of standards and bring order and interoperability to the mayhem.

Matter's official website defines it as the following: "This industry-unifying standard is a promise of reliable, secure connectivity – a seal of approval that devices will work seamlessly together, today and tomorrow. Matter is creating more connections between more objects, simplifying development for manufacturers and increasing compatibility for consumers. This collaborative breakthrough is built on proven technologies and guided by the Connectivity Standards Alliance (formerly Zigbee Alliance), whose members come together from across industries to transform the future of connectivity." Many smart home devices use proprietary protocols today, requiring them to be tethered to a home network using dedicated proxies and translators. By building upon IP, some of these devices may instead be able to connect directly with standardized networking equipment (Figure 3.6). This is a massive effort. As of 2021 more than 200 companies have signed up to participate in Matter. The list includes several of the biggest brands in the industry. The devices that are listed as focal points for the alliance include lighting, HVAC systems, smart locks, security systems, and shades. Interestingly smart speakers or large appliances have not been included in the effort. In late 2020, the alliance announced that the goals of the project also include commercial IoT use cases that include offices and other nonresidential buildings.

Currently all the open-source design documents and the code is publicly available in the project's GitHub repositories. Figure 3.7 zooms in on the various layers of the standard. The open-source code intends to leverage existing capabilities in the following list among others:

- Amazon's Alexa Smart Home
- Apple's HomeKit

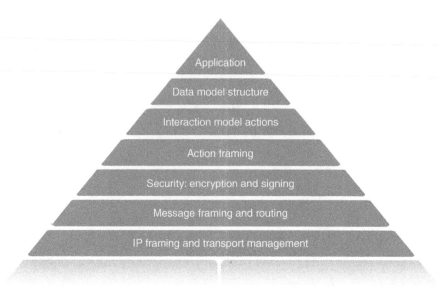

Figure 3.7 Matter's data architecture diagram. (*Source:* Image courtesy of Matter; Project Connected Home over IP [CHIP]).

- Google's Weave
- Connectivity Standards Alliance's Dotdot data models

The goal of the first specification release will be Wi-Fi, up to and including 802.11ax (also known as Wi-Fi 6), that is 802.11a/b/g/n/ac/ax; Thread over 802.15.4-2006 at 2.4 GHz; and IP implementations for Bluetooth Low Energy, versions 4.1, 4.2, and 5.0 for the network and physical wireless protocols. Matter will likely also embrace other IP-bearing technologies like Ethernet, Cellular, Broadband, and others over time. While some companies might focus their product offerings on the protocol over Wi-Fi or Ethernet, others may target the protocol over Thread or BLE (Bluetooth Low Energy) or a combination. By specifying acceptable IP-based protocols under one big umbrella, manufacturers and consumers will be able to build and buy with an increased peace of mind.

We expect that over the coming years this alliance and published standards become more pervasive as a result of its benefits and the backing of the largest players. Consumers will like it because they can mix and match their IoT gadgets; as long as they are Matter-certified, they will interoperate. This is in contrast to the pre-Matter world where an Apple Homekit would not talk to a Google Smart Speaker. The architectural diagram of the standard is quite rich (Figure 3.8). We bring it here for reference courtesy of Matter.

Example: Improve Quality of Life and Energy Efficiency by Smart Thermostats

Smart thermostats have been around for several years now. Nest, which was later acquired by Google, was one of the more popular brand names in this market bringing smartness, function, and style all together. There are several other brands including the ecobee line

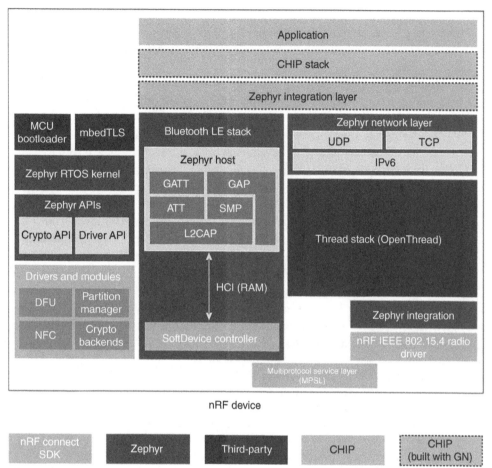

Figure 3.8 Matter's architectural diagram.

from Amazon. In coming up with their selling message most of these companies allude to a few standard factors as their marketing message:

Cost saving: in one of Google's support pages, they say that "On average the Nest thermostat saved US customers about 10–12% on their heating bills and about 15% on their cooling bills. We've estimated average savings of $131–$145 a year, which means the Nest thermostat can pay for itself in under two years." This message is pretty simple on the financial side. It also helps sell to the climate-savvy buyers who care about their carbon footprint.

Comfort: smart thermostats do a number of things that improve the comfort level of users' houses. On the one hand, they learn the occupants' routines over time and make themselves attuned to such temperature and timing habits. They also allow firing of heaters or ACs from afar over the Internet for a warm enough or cool enough home to get back to. Some models also allow automated activation of the heating and cooling systems based on the location of the occupants. The location is read through the smartphone that is carried by the user, which is a textbook example of how things connect over the Internet to optimize a task.

Convenience of use: most smart thermostats integrate well with digital assistants so that temperature setting can be adjusted using a human's voice. The over-the-net control feature also allows forgetful people to turn off their heaters manually if they are on an extended leave.

Style: we have come a long way since the typical boxy and edgy thermostats of the twentieth century. Newer smart thermostats offer a stylish and well-rounded design that is also easy to operate.

The data that is being collected by the Nest thermostat is optionally used by Google to monitor the performance of the devices they sell you. They also allow themselves, with the user's permission, to use the data for direct marketing purposes. That is an extremely valuable by-product of this smart home gadget that opens new business models to the manufacturer. For example, knowing that someone's home temperature fluctuates significantly during the colder months of the year guides the marketer to show them products and services that help make a home more sealed and insulated. Or knowing that a home's temperature pattern frequently fits the "unoccupied pattern" during winter, you can assume that the occupants are into winter vacations or that this is their second home. The marketing message to the occupants can be fine-tuned based on this extra piece of knowledge. Of course, there are strong privacy considerations in this scenario. In order to get a better idea of what kind of data Nest thermostat gathers, here is an excerpt directly taken from Google's support pages:

What Data Does My Nest Data Provide?
Your data archive from My Nest Data includes information you've provided about yourself, like your email address; details you've provided about [sic. and] your home, like your address and room names you've assigned in the Nest app; and sensor data from Nest thermostats, smoke alarms, or cameras you've connected to your Nest account. It also includes information about your device interactions and app usage. If you have a Nest Aware subscription, and you requested your video history, we'll send you that, too.

We will see more about the peripheral business opportunities that consumer data can offer in the next segment. However, such uses need to be considered with utmost care and in close consultation with legal and public relations professionals. Handling users' data with care and respect pays off over time even if short-term attractions are tempting. Not only do rules change over time, they also vary significantly from one country (or even state) to another. One simple misstep can cause massive legal, financial, or public relations costs. In our example, Google explains their terms of use in the following way to clearly define the matter while providing some degree of control to the consumer over their data:

Does Nest Sell the Data That It Collects from My Home?
Nest does not share personal information for any commercial or marketing purpose unrelated to the activation and delivery of Nest products and services without asking you first. We may use information you provide to send you surveys about your Nest use or other direct marketing and communications from Nest. You can configure your marketing preferences in the Nest app or at nest.com.

CONNECTIVITY CHOICE AND ITS IMPACT ON ROI

In industrial and business use cases, the choice of wireless or RF connectivity is of paramount importance. It is not to say that such a decision is trivial in consumer applications; however, measuring the return on investment for industrial scale IoT applications makes this decision of special importance. While in consumer use cases, ease of set up, reliability, user-friendliness, and compliance with regulatory requirements are the more important factors, in industrial use cases, the calculation may be somewhat more involved. Like most other decisions we need to make sure that the design decisions make business sense. In this segment, we briefly look at the RF connectivity technology and its role in the cost of an IoT network. Such cost should then be compared with the benefits of a connected system to determine viability.

In particular, we will look at the choice of connection technology. While this is only one of the decision factors, it will impact several other aspects of the network and overall cost. The immediate impact of a specific RF connectivity and networking architecture is the capital expenditure (CAPEX) required to implement the network. Longer range, higher reliability, more modern standards, slicker designs, and similar factors contribute to the investment cost. However, the cost to run the network over time is extremely important as well and should not be overlooked. While an inexpensive design can be up and running quickly and cheaply, over time its maintenance cost can easily outweigh the initial savings. In what follows, we review some of these decision factors both from initial CAPEX and ongoing costs.

When it comes to the upfront cost, a good process usually starts from plug and play and available modules in the market instead of custom design. This is extremely important because overdoing a custom design too early can drastically elongate the design and test process unnecessarily, thereby adding to the implementation and opportunity costs. Many connectivity, analytics, microcontrollers, PCB boards, and similar modules exist in the market that can expedite the initial process. That way, the higher levels of network and system design can be tried and tested without getting bogged down with the last mile optimization or customizations. Make sure that a prototyping phase remains a prototyping phase and not creep into too deep customization.

Another aspect of the networking design is the architecture. The two common choices are mesh or star architectures. While the former allows for shorter range communication to be viable it adds extra nodes and repeaters to the topology and makes the protocols more sophisticated. A star topology, on the other hand, allows for fewer nodes and simpler protocols at the cost of longer range communication that may require more powerful individual nodes. As we will see shortly, these choices will have ongoing maintenance cost implementations as well that need to be taken into account.

The choice of hardware for various modules in the system is both a driver and a function of other design choices. Shorter range communication protocols call for more hardware nodes and potentially beefier processors to act as repeaters in a mesh network. Longer range designs can work with fewer hardware modules but will require longer range and stronger

RF modules to be incorporated. Aside from such specifics sticking with interoperable and modular pieces of hardware can help reduce cost both initially and over time. While choosing one custom monolithic system to cover end to end may sound very appealing in the beginning, it can cost more to buy such a system of high customization. Perhaps more importantly it limits the optimization opportunities that can come from picking and choosing various modules based on their economy. Over time, the problem can and will get exacerbated because it locks the buyer with the initial vendor. The technological ramification of this lesson is to pick modules, chips, antennas, and alike in an interoperable way and according to accepted and widely adopted standards as the first choice.

The first time roll out of an IIoT system can be significantly costly and the choice of communication system plays an important role in it. A key aspect is whether or not the distributed sensors and actuators will require wired power. Certain communication choices such as Wi-Fi will practically require hard wired power to the nodes because of how much power they consume. Battery-operated nodes, on the other hand, can be installed in various places much more flexibly and thereby reduce the upfront cost. The networking protocol and design should also be done in a way that integrates with the existing networking infrastructure to avoid extra cost.

By now it should be no surprise how the connectivity choices impact the ongoing costs of running an IIoT system. Network management and device maintenance can be costly as well. Tasks such as troubleshooting, network security, report generation, provisioning and offlining nodes, and similar administrative operations require dedicated resources and skills. Battery-operated designs can help with the flexibility of the network design by offering many more options for the nodes' physical locations. However, the choice of the connectivity design and networking protocols should be friendly to the limited amount of energy in batteries. Frequent replacement of batteries will incur both human and equipment costs over time.

While in-house networks will need upfront design and integration, commercially available cellular and satellite networks will add ongoing cost. More integrated environments tend to be more suitable for managed networks, while more distributed industrial or environmental use cases can leverage commercially available wide area networks such as cellular or satellite services. In the latter case, it is important to have a good idea about the communication needs of each sensor and use that to calculate the cost. For example, satellite communication is often suitable for very short bursts of messages that are not sensitive to delay. Many IIoT sensors can leverage lightweight protocols such as MQTT to minimize the size of their communication bits while saving energy.

CLIMATE CHANGE AND PERFORMANCE PER WATT

Moore's Law has been the ruling law for silicon development since the 1960s. Gordon Moore, the co-founder and ex CEO of Intel, predicted that the number of transistors per unit of area or chip would double every year. Over time, it slowed down to doubling every 18–24 months. Nevertheless, the prediction remained quite accurate. As a result the industry moved from 2250 transistors in an area of 12 mm^2 to more than a hundred million transistors per square millimeter. This development meant that the same circuit designs just automatically run more efficiently over time without any significant changes to the design itself. That is until the trend slowed down recently due to physical limitations of the materials used in manufacturing chips. On top of that, even when the size of transistors became smaller after a certain point the resulting circuit would not provide the same power efficiency that the earlier leaps in shrinking provided; the low hanging fruit was gone.

In 2010, Jonathan Koomey, a Stanford professor, coined a new law. Koomey's law trends the number of computations per joule of dissipated energy. Per the law, this index has doubled every 18 months from 1945 to 2000 (100× per decade) and then slowed to doubling every 2.6 years or so since (16× per decade). Figure 3.9 shows the concept of Koomey's Law.

Given the state of our planet and climate change, Koomey's law is ever more important. Designers need to pay attention to not only the size of the circuitry but also the energy footprint of the design for the task at hand. For example, an Application Specific Integrated Circuit (ASIC) that does one and only one job efficiently can in certain cases be a superior design choice than a powerful general purpose CPU that offers more flexibility and unused horsepower while burning more power. The former can even be designed in older chip technology (e.g. MOSFET of 130 nm channel length as opposed to the modern 3 nm channel length) and still provide good economy. Note that such older silicon technologies require less niche-skilled designers and modern fabs to build, thereby delivering a good economy of production in other ways. Of course, if the product design requires a versatile CPU, then such a choice would be inevitable and the wise way to go. The main point of the argument is that the calculation of the economy should be beyond just Moore's Law. A good IoT designer should consider the full energy footprint of their design and potentially find a happy medium between size, energy consumption, and flexibility.

A direct corollary of this argument is that the computational algorithm can play a significant role in the energy footprint of the devices. As more computational power is added at the edge of the IoT network with embedded artificial intelligence (e.g. for face recognition) squeezing the algorithms to exactly what's needed can have a big impact on the energy consumption. For example, a neural network with millions of nodes that deliver only marginal improvement over a much smaller decision tree may not be an economically wise choice.

DATA BY-PRODUCTS BEYOND THE ORIGINAL IOT APPLICATION

In this segment, we will discuss a number of different business models that can arise from the IoT applications beyond the original intent of the product. Note that *beyond the original intent* is quite relative; depending on what the original intent has been, some of the models

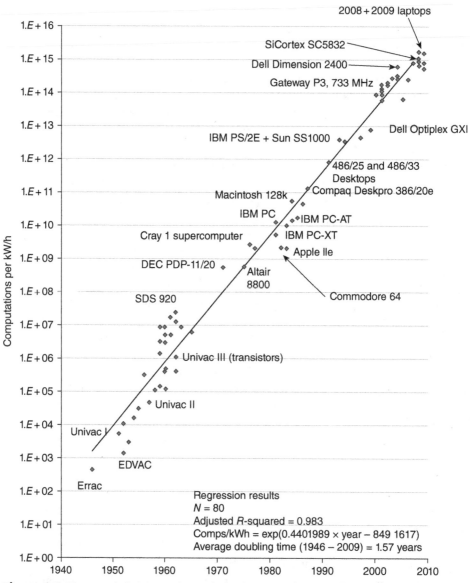

Figure 3.9 Koomey's Law graph made by Koomey (*Source:* https://en.wikipedia. org/wiki/Koomey%27s_law).

may be considered extensions of the original intent while for another application the same idea may be considered an additional business model. Regardless we will enumerate several business possibilities that an IoT product can enable. While taking advantage of some of the reasonable opportunities is highly encouraged, we like to warn against wanting to become everything. Enabling any one of the following business models is a serious undertaking that will consume resources. So, don't forget that focus matters a lot.

Services and Device Management

Managed industrial or consumer services: IoT enables certain services to be performed remotely. As a result it may make economic sense to outsource them to an expert partner and pay them for such a service. As a result the company that receives the service can reduce cost by avoiding hiring experts in house, scale their operations up or down much more easily, and focus on their core business instead of trying to become experts in the machinery. As an example, imagine an investor business that buys a wind farm to produce electricity and produce profit for their shareholders. Their core business is investment in a diversified portfolio and to maximize the profit. They are not experts on how a wind farm is run. IoT-enabled devices help them outsource the management of their wind turbines to external expert companies. The service provider takes advantage of the economy of scale by offering its services to multiple customers and can lower the per-unit cost. This is a model that is already in place in several countries in the world with advances in renewable energy.

Management of end nodes: like any other electronic or electric device, IoT devices need to be managed. In the most common scenario, new devices join an IoT network or leave it regularly. Therefore, new sets of resources, communication channels, and identities should be provisioned on a regular basis and according to certain rules. The software on existing nodes needs to be updated and patched from time to time. All such end node management can be offered as a service to users who may not have the expertise or bandwidth or intention to do this at scale.

Real-time analytics: beyond collecting data, performing analytics in the right context is something of tremendous added value. For example, if a pump starts jittering in a way that is known to be a sign of wear and tear, it may be taken down gracefully and fixed. If a new face is recognized inside a house while nobody else is at home, a notification needs to be fired. In these examples and many more, analytics is done close to real time leading to notifications or other advanced processes. The analytics can be performed by external providers who gain access to the data through IoT networks.

Peripheral Businesses

All of the above through data access: we talked about a number of potential services above. Now the question is "Who should do that?" Remember that building and selling an IoT system may be a very different kind of business from offering such services as anomaly detection or servicing equipment. While the IoT vendor can directly offer such services, it can also be done through partners. An example of the former model is when a smart camera vendor also offers detecting anomalies such as motion or new faces and sending notifications. An example of the latter is when the IoT company grants a partner access to the data from all pumps on the customer's factory floor so that the partner can offer their service to the end customer. The sensors gather the data from all the pumps on the shop floor at scale and store them in a data warehouse in the cloud. The partner companies can pipe the data to their own engine for their analytical service. Such arrangements offer a lucrative expansion to the revenue while enriching the ecosystem of products and services. That in turn creates more gravity for the IoT products. The cost of making these extended services happen is the extra engineering and process efforts that are required to provide customer data to the partners. Granting permissions to the customer data by third-party organizations also has to be taken with utmost seriousness. Not only should the customer be aware and accept every individual access, the parent IoT company should also trust the partner sufficiently. Any malfunction or data breach by a partner can be disastrous for the parent IoT company.

Selling data to utilities and insurance: a special form of peripheral opportunities is gaining quite a bit of traction especially in the consumer market. Meter or submeter electricity data from a house, operational details from a car, or water consumption details from a garden can all be valuable to utility and insurance companies. That's why many such companies persuade consumers directly or indirectly through cheap IoT devices to measure and share their data. In return, they usually offer inexpensive smart devices, discounted or customized services, or just cash. In return, they ask for access to the data so they can understand their users' behavior much more closely and make their pricing and marketing campaigns more targeted.

Marketing

Having access to customers' data enables marketers to hyper-personalize their messages. For example, most consumer IoT products come with end-user applications or web portals. In those applications or portals, the IoT vendor can market services or products that are more likely to be of interest to the targeted customer based on their situation. A customer who may be experiencing a leakage in their faucet is likely to be receptive to an ad for a plumbing service. Or an industrial user who experiences a significant increase in the downtime of their pumps may be interested in a promotional offer by a consultant to investigate the cause. Such marketing can be done directly within the user experience with the product or outside through data access. Our application or portal example was one of the former scenarios. In the latter case, an advertiser, such as Google, can leverage the consumer data to boost the quality of the ads.

Data Management

IoT customers generate and own an increasing volume of data in the form of measurements, events, photos, video snippets, health measurements, or other custom forms. Over time these data sets become both more valuable and harder to manage at the same time. Not only the size of the data set grows, finding episodes of interest becomes ever more challenging. Handling and managing such data sets at scale securely together with added services such as fast contextual search or community sharing is a common business model by many IoT companies. Many home security companies offer management of security video captures in the cloud for a nominal subscription fee.

Analytics and AI

We talked about real-time analytics under *Services and Device Management* above, so we won't repeat that here. There are other offline or batch types of analytics that can be offered to customers. We use the word analytics quite broadly here to include simpler statistics of the data, as well as more advanced operations such as machine learning and AI. When the volume of data increases and other similar data sets become accessible to the IoT company, suddenly they are empowered to do much more with them. Data sets covering longer periods of time enable analytical services to inform the customer of their inefficiencies, trends, predictions, anomalies over time, among other simpler observations. For example, some smart thermostats learn the users' behavior over time by applying analytical algorithms to the behavioral data set. After a number of months of measurement, a facilities company can inform office managers of the inefficiencies in their large office buildings due to heat leakage. As we see below, these real data sets can also be offered to other developers in an anonymized fashion for their product design and development. Transferability of data and analytical models boost the economy of data exponentially.

Developers

A growing market is opening for smart products and services in the world. Many such business models revolve around a central machine learning algorithm that makes predictions, recommendations, or deductions based on data. These models need to be trained with data. The volume and quality of the training data set is extremely important to the quality of the final product. Besides, in most cases the data needs to be labeled. Real-world data gathered by IoT devices can bridge that gap. In fact offering data as a product is an attractive offer for many developers. While these data sets can be sold for direct profit in many cases, the IoT companies use them to lure developers into their development ecosystem. As a developer you would want to build an application where real data exists for your training or design purposes. Therefore, offering such data sets exclusively to developer members creates a significant pull for companies and a sense of exclusivity. Needless to say, such offerings should be handled with a tremendous amount of care. Not only individual customers should be aware how their data is being used, local laws clearly define and limit such use cases in many countries. Anonymization or obfuscation of the data is sometimes a mitigating factor. On the flip side any step that makes the data set less like its original shape reduces its value. So it is recommended that organizations find the right balance between caution and total recklessness when designing their business model with customer data.

Security and Safety

Real-time notifications: a by-product of many IoT applications can be detecting security and safety incidents. For example, if a consumer smart plug detects a larger than usual load from a lamp, they can notify the user of a potential safety risk. If a smart garage door opener detects a door opening at a time that is highly uncommon, they can let the consumer know of a potential breach. An industrial temperature sensor that is used to regulate and optimize energy usage in a factory can notify the managers of potential overheating of devices or fire.

Trend analysis: extending the use cases above is when historical data is used in batch to detect general trends and notify drifts and deviations. A smart air quality sensor can detect a gradual deterioration of the air quality inside a room, which may mean that a filter needs to be replaced in the AC. This is indeed a safety use case.

Sharing among a community of users: in both cases above, IoT companies can act as an information hub for their users to add value while helping their users to be safer. Some smart doorbells and home security systems let users share such events as burglary or wild animal sightings in a neighborhood. This kind of access can be offered as a value add for a subscription fee or just included for a better service experience. Industrial service providers who use the same type of equipment can be notified of incidents when their assets run in a state that is far from the average user in the community of users. The IoT company can build the benchmark operational signature based on its universal observations and offer security and safety services based on deviations from those global benchmarks.

Chapter 4

Security

There is no state of absolute security. Nevertheless, attempting to reach a state of acceptable security for any data and software system is an absolute must. Data privacy is a close cousin of cybersecurity that is also extremely important. A secure system understands which user is attempting to take certain actions with the data, hardware, software, or processes and applies the predefined authorization rules accordingly. Therefore, only legitimate users who are supposed to take actions in the system will be granted such privilege. Data privacy, which has become increasingly important over the past few years, defines principles by which entities can have access to users' data. Those include intentional and unintentional access to the data such as marketing or accidental breach of user's personally identifiable indicators. Loose or weak levels of security cannot guarantee a reasonable level of data privacy. There are numerous examples where poor choices in security have led to tremendous amounts of loss or brand damage to companies. The issue has become important enough that a large behemoth such as Apple often touts its strict considerations for data privacy as a major differentiator in the consumer market. On the other hand, breaches in systems such as Verkada cameras in 2021 have led to massive headaches for the executives while inviting the FBI to investigate.

In this chapter, we will spend a significant amount of time going through several security principles and methodologies. The goal is not to make the reader an expert in security. Rather it will give the reader a high-level view and great understanding of what those principles are, what terminology to use, what questions to ask, and most importantly how to think about security as a principle.

To begin, you need to consider security as a way of designing and thinking as opposed to a clearly defined state of a system. Although there are several standards and tests that give you a binary pass or fail result, those results are often proxy measures for a bigger

IoT Product Design and Development: Best Practices for Industrial, Consumer, and Business Applications, First Edition. Ahmad Fattahi.

goal. Given how often hardware and software systems change, secure processes of today may become more vulnerable tomorrow. There is an uncountable set of examples of new chips or software patches that have been shown to open new security holes. Besides, systems are only as hardened as the imagination of their designers. Hackers with deep pockets (governments), lots of time and energy (the youth), or iron will power (competitive hackers) are constantly thinking outside the box to break into systems.

The motivation behind hacking can often be financial. Hackers gain control over assets that they sell on the dark web, or flat out demand ransom from the victims in order to not release the sensitive information. In some other cases, the motivation is gaining a geopolitical edge over opponents. Many countries are changing the rules of warfare by taking the attack surface to the Internet and cyberwarfare; it's much less expensive and also much harder to prove liability. Therefore, it has been common for adversarial or competitive countries to attack other countries' critical infrastructure control systems (for example utility or water) or military systems. Some other hackers break into systems mainly to prove a point or gain others' respect. This is more common among younger attackers who are part of a circle of hackers. Whatever the motivation, the outcome for the victims can be massively disastrous. It can lead to financial loss, damaged reputation, losing competitive edge, or legal costs.

A common misconception among executives and system designers is beliefs such as *we have built a fully secured system* or that *we have tested every possible attack and blocked the penetration path*. To be successful in the realm of system security, you need to think like an attacker. It means you should always assume there are ways beyond what you could have imagined. It also means that security is a relative measure. As we go over several concepts and security methodologies in this chapter, you need to interpret them as ways that make the hackers' job increasingly difficult. Interpreting them as absolute security is wrong. Successful security professionals and hackers often think outside the box. They tend to think *orthogonal* to how most other people think. They are contrarians who keep asking "what if" questions. They beat up systems in various less obvious ways to see what might happen. That's why your level of security is as good as the imagination of your security designers.

A corollary of this phenomenon is that it is almost always recommended to use established algorithms as opposed to getting creative with building your own security systems from scratch. Standard security algorithms have been beaten up for a good amount of time by many smart people. Only the reliable ones have been passed to the level of becoming a standard. Therefore, it is safe to assume breaking them by a hacker should be very hard or should take lots of time. It may be very tempting to come up with your own novel algorithm that no one has tested before. However, the chances of missing big holes or scenarios go significantly up. If, for any reason, you decide to rely on an in-house algorithm at least include several people with different backgrounds in the process. Let the tinkerers and contrarians in your organization play with it and try to break it. Set them free and encourage them to use any physical, software, or even social engineered way to break the algorithm. The cost you incur up front in this process would be dwarfed by a malicious attack after the algorithm ends up in your products.

Incident response is another extremely important aspect of best security practices. Given the nature of security attacks it is only reasonable to assume that breaches are bound to happen; it is a matter of when and not if. Once such incidents are discovered, depending on the nature of the incident, timely and clear communication is of high importance. Prudent teams envision several different scenarios, define chain of command in response and communication, run drills, and constantly update themselves. Sometimes a well-designed response and communication strategy can save a significant amount of loss that would otherwise accrue.

Kerckhoffs's principle states that a cryptographic system should be secured even if everything about the system, except the decryption key, is public knowledge. It means that you should not assume that the design details of your system will remain secret. Security by obscurity, in contrast, relies heavily on keeping design details secret. It, therefore, creates security through secrecy. The principle, although useful, has been generally ruled out as the only way to make a system secure. For example, a saying in the US National Security Agency goes like this: "serial number 1 of any new device goes to the Kremlin." While shrewd companies don't advertise their sensitive designs, they also don't rely solely on its secrecy. In some recent IoT incidents, hackers were able to reverse engineer where network passwords are being stored on chips in plain text. That's a perfect example of violating Kerckhoffs's principle. In what follows in this chapter, we will review several methods to create higher levels of security through establishing encryption keys.

ENCRYPTION TECHNIQUES

In this segment, we review a number of popular encryption techniques and discuss their strengths and weaknesses. As a reminder, the goal of this segment is to familiarize the reader with the main concepts. There is a rich set of literature on the topic that security experts need to study before designing a system. Government bodies and professional consortiums publish many standards and best practices on a regular basis.

Let's start with some basic definitions. Plaintext is the actual unencrypted content that we like to protect from hackers. Ciphertext, on the other hand, is what comes after encrypting plaintext, which is often done by using a key. Another concept that we will introduce is the notion of symmetric versus asymmetric algorithms. While in the former the same key is used for encryption and decryption, the latter uses two different keys. A threat vector or attack surface is defined as the ways through which data can be leaked in an unintended manner. In reality, threat vectors cannot be reduced to zero. However, the goal of security standards and algorithms is to make such threat vectors as small as possible.

Symmetric Encryption Algorithms

We will start by introducing a number of symmetric algorithms such as Caesar cipher, One Time Pad, Advanced Encryption Standard, and a number of its variants.

Caesar Cipher: Caesar cipher or Caesar shift simply shifts every character by a number of letters to left or right. For example, replacing A with C, B with D, and so forth is Caesar ciphering by 2. As easy as it is to understand, it is also very easy to crack by hackers. The technique's namesake is Julius Caesar who is said to have used this technique to communicate sensitive messages. As you can imagine, even though it was effective at the time, it does not provide much security today. If the attacker knows that the technique is used but doesn't know the shift amount, they can easily try a small and finite number of integer numbers as the shift amount to see which one breaks the code. If they don't know for sure that the technique is used, they can apply frequency techniques and start forming hypotheses quickly about common words such as "the," "a," "and," "or," and others. From there they would be back to the previous scenario.

One Time Pad: Another simple technique is called One Time Pad (OTP) also known as perfect encryption. It assigns a number to each letter ($A = 0, B = 1, C = 2, \ldots Z = 25$). There is a key that is mutually decided and is added to the actual text. The key should be at least as long as the message itself or gets repeated according to a prior agreement between the two sides. The plaintext and key are added together and the result is calculated modulus 26 to create the encrypted message. The reverse calculation is done at the receiving side to decrypt. The key is agreed upon in advance between the two sides. It can be something like the following: agree on a specific random book in advance and in secret. Then on every day of the year, take the page of the book corresponding to the day; the string of characters starting on that page constitutes the key of the day. The technique is very hard to crack but also impractical in the real world. Scaling it to more than a few channels a day becomes a real problem. In comparison, imagine that every single web user or IoT device establishes several encrypted secure sessions with different parties on any day. Figure 4.1 shows how the message HELLO gets through a communication channel using the key value XMCKL.

Advanced Encryption Standard: Advanced Encryption Standard (AES) was originally established by the United States National Institute of Standards (US NIST) in 2001. It has a wide range of applicability in web browsers, wireless communication, processor security, and file security. AES is a subset of the Rijndael block cipher algorithms that was developed by two Belgian cryptographers, Vincent Rijmen, and Joan Daemen. They submitted a proposal to NIST during the AES selection process. It is a symmetric

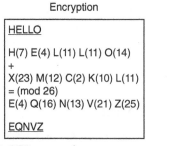
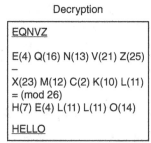

Figure 4.1 OTP example.

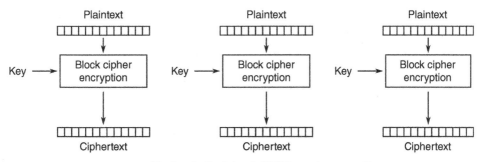

Electronic Codebook (ECB) mode encryption

Figure 4.2 ECB encryption (*Source:* https://en.wikipedia.org/wiki/Block_cipher_mode_of_operation#).

algorithm, meaning the same key is used for encryption and decryption. The technique encrypts blocks of data, one at a time, with an encryption key of the same length as the data block (Figure 4.2). NIST picked three members of the whole decryption family and standardized them: the block length is 128 bits while the key length can be 128, 192, or 256 bits. First an Initialization Vector (IV) is used to generate a number of subsequent keys. All of the produced keys will sequentially encrypt blocks of data a fixed number of times. The version with key length 256 is virtually impossible to crack because of the time it would take to try a brute force attack. The US government, for example, approved all three key lengths for confidential government content up to the SECRET level. TOP SECRET requires the 192- or 256-bit key. In the following few paragraphs, we will review a number of different implementations of AES and review their pros and cons. The main takeaway to draw is awareness about these various flavors while gaining an understanding of the factors that make an algorithm valuable for a certain use case.

Electronic Codebook (ECB) is one of the simplest forms of AES. In this algorithm, the entire message is broken up into blocks of appropriate lengths and encrypted separately using the key.

The advantage of this scheme is that the encryption and decryption can be parallelized for fast operations. It also offers random access to any part of the message without forcing a full decryption. However, a massive drawback of it is that it "leaks" information through preserving certain patterns in the data. A great example of the information leakage is shown in Figure 4.3: while trying to encrypt the Linux logo, the top row shows the effect of the encryption with ECB, while the patterns are obviously preserved to the level that can be guessed easily. This phenomenon is something that is a significant concern for any encryption algorithm. For comparison, compare that with the row below it. The bottom row creates a pseudorandomness that is much more desired in an encryption scheme.

Cipher Block Chaining (CBC) tackles this major drawback by combining (XORing) the result of the last encryption block with the unencrypted data of the next block (bottom image of Linux logo). Figure 4.4 shows how the blocks connect with each other. The cost of this scheme over the simpler EBC is that the encryption algorithm is not parallelizable

Figure 4.3 ECB leaky encryption.

due to the dependency of the next block on the result of the previous block. However, it still enjoys parallelizable decryption and random access to the data. So, if the source is not time sensitive, the cost is not a big concern.

Propagating Cipher Block Chaining (PCBC) takes things one step further: each block gets XORed with both the input and output of the previous block before getting encrypted (Figure 4.5). The result is more randomness in the encrypted output. However, we lose parallelizability for both encryption and decryption, as well as random read access to the data in the encrypted version.

In all of these modes, the Initialization Vector, or IV, should be chosen wisely in a way that is unpredictable and unique. Remember to think like a hacker: good hackers try obvious choices such as a company's name, user's date of birth, lazy sequences such as "qwerty" or "12345" or similar options.

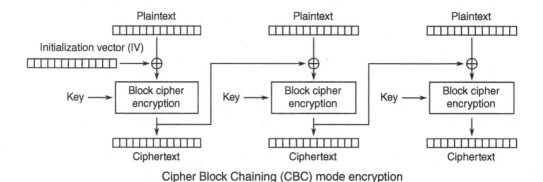

Cipher Block Chaining (CBC) mode encryption

Figure 4.4 CBC encryption (*Source:* https://en.wikipedia.org/wiki/Block_cipher_mode_of_operation#/media).

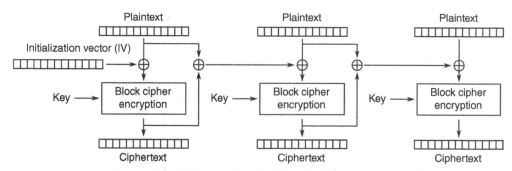

Propagating Cipher Block Chaining (PCBC) mode encryption

Figure 4.5 PCBC encryption (*Source:* https://en.wikipedia.org/wiki/Block_cipher_mode_of_operation#/media).

As you can imagine there are several other variations of these block cipher algorithms such as Output Feedback (OFB). OFB is a stream cypher algorithm where plaintext digits are encrypted with a pseudorandom cipher digit stream (keystream). In a stream cipher, each plaintext digit is encrypted one at a time with the corresponding digit of the keystream to produce a digit of the ciphertext stream. The Counter (CTR) mode enables encryption on a stream of data where a counter plays the crucial role of initiating the next keystream. These stream algorithms work well on streaming data and also show more resiliency to occasional errors in the unencrypted data. And finally the AES XTS mode (NIST SP800-38E specification) is mainly used for encrypting storage devices.

What remains common among these modes and flavors is the existence of an Initializing Vector of a certain length that gets combined with blocks of equal size of the unencrypted data. How the combining operation happens defines different versions of this algorithm. The important factors to keep in mind are below. The choice of the algorithm should eventually be done by making the right prioritization and compromise among the following factors depending on the business and engineering requirements:

- The level of protection (how difficult is it to break)
- How much information can leak
- How expensive the encryption is
- How expensive the decryption is
- Is random access to plaintext possible during decryption
- Is encryption parallelizable
- Is decryption parallelizable
- How error propagates in the data

Error propagation deserves a short description. Imagine that due to physical or malicious reasons, there is one bit of error in the unencrypted data; how does this error propagate through the rest of the data? In stream cipher modes such as OFB and CTR, such errors can be tolerated much more easily. That is because the whole process is transient. However, other modes such as ECB and CBC will reflect the result of a bit error in one or two blocks due to the dependency of future blocks on previous blocks.

Asymmetric Encryption Algorithms

AES is an example of the encryption algorithms known as symmetric encryption; the same key is used to encrypt and decrypt data. There are also a whole host of asymmetric encryption algorithms widely used in the industry. These algorithms are sometimes referred to as *Public Key Encryption* because they use a public key, as well as a private key. The two keys are different but mathematically connected depending on the algorithm of choice. It's like a lock with two different keys: one key can exclusively lock while the other one can exclusively unlock; having the key that locks doesn't really tell you much about the shape of the key that unlocks. The reason these techniques were invented is to scale the problem of creating unique and secure connections between two parties such as a server and a client or two peers in the network. One party can publicly reveal a key by which other parties in the insecure world should encrypt their messages. While each of these parties encrypt their messages using that publicly known key, a different (and hard to guess) key should be used to decrypt the message. That is the private key that is kept secret by the original publisher of the public key. That way, using *one-way mathematical functions*, two related keys can be used to secure a channel: while one can be safely published for anyone's seeing, the other one is kept secret. The asymmetric algorithms can be used for both encryption and authentication as we will see later in this chapter. It means, not only the data itself becomes protected, the sender of the message gets verified as well. Asymmetric algorithms are often slower than symmetric ones. Sometimes when speed is an issue, a server and client establish a symmetrically encrypted channel (which is faster) through an initial asymmetric process. The main challenge in establishing a symmetric encryption securely is sharing the common key securely. The server can do that at the beginning of a session through asymmetric encryption. The symmetric key can be encrypted by the server using the client's public key and transmitted securely to the client. The client, and only the client, can then decrypt and see the shared symmetric key that is sent by the other party; from that point on the two sides can communicate securely using the established common symmetric key.

Asymmetric algorithms use two keys where one gets communicated from one party to the other and is used to initially encrypt the data. The unauthorized listener can potentially listen to the channel and record the public key. However, the private key is needed to decrypt the data (hence the notion of one-way functions). Therefore, even if a hacker eavesdrops on the communication channel and has the public key in their hand, they still cannot make sense of the data. It is also noteworthy that even though public and private keys are mathematically related, going from the public key to the private key is practically very hard and computationally expensive. See Figure 4.6 for a simple schematic of an asymmetric encryption process.

Figure 4.6 Asymmetric encryption.

Now the question may be why we don't use asymmetric encryption all the time if it has the extra security? The answer is that this extra layer of security and handshake between the source and the destination makes the process slower. A familiar example of both of these algorithms is any HTTPS website. In the initialization phase, where the source and destination need to authenticate and make sure both are who they say they are, an asymmetric public-key encryption method is used. Once the trust is established, symmetric algorithms are used for faster and more efficient communication of the actual data.

Diffie–Hellman Key exchange: in 1976 Whitfield Diffie and Martin Hellman proposed a protocol for two parties to establish a common secret (key) over an insecure and public channel. This protocol builds on the work of Ralph Merkle who proposed some of the underlying ideas in late 1960s; hence, in 2002 Hellman suggested that the protocol be called Diffie-Hellman-Merkle. The main question that is being answered by this protocol is this: the two parties who are to establish an encrypted communication need to establish a secret key between them. As we saw, earlier algorithms such as One Time Pad (OTP) do a great job at this as long as the two parties can agree on the "key of the day" in advance. In real-world IoT and Internet scenarios, billions of new communications need to be established every day. Therefore, the initial part of agreeing on a common yet secret key needs to somehow happen out in the open over an insecure channel. That's where the Diffie-Hellman algorithm comes in.

Let's walk through an example. The analogy we will use is two different shades of paint. The main idea is that mixing two shades of paint together to get a new shade is much easier than separating the result back into the original shades. In other words, the algorithm devises a one-way mapping where locking is done computationally easily while unlocking remains computationally impractical. Let's consider a numerical example. Assume that Alice and Bob are trying to establish their secret key while Eve is eavesdropping on them. In the text below, we show secret values that only Alice or Bob know in bold and underlined font. Every other value can be eavesdropped by Eve.

1. The first step for Alice and Bob is to agree on two integers publicly, which can be proposed by one side for practical reasons: the first number p is a prime number and the second is g that is a primitive root modulo p (which means for every integer m coprime to p, there is some integer k for which $g^k \equiv m \ (mod \ p)$).

2. The next step is for Alice to pick a secret integer $\underline{\boldsymbol{a}}$ and for Bob to pick a secret integer $\underline{\boldsymbol{b}}$. These two integers never go across the insecure channel.

3. Alice sends Bob $A = g^{\underline{a}} \bmod p$ and Bob sends Alice $B = g^{\underline{b}} \bmod p$.

4. Alice computes $\underline{\boldsymbol{s}} = B^{\underline{a}} \ mod \ p$ and Bob computes $\underline{\boldsymbol{s}} = A^{\underline{b}} \ mod \ p$.

Table 4.1 An example of Diffie–Hellman Key exchange.

Alice		Bob		Eve	
Known	*Unknown*	*Known*	*Unknown*	*Known*	*Unknown*
$p = 23$		$p = 23$		$p = 23$	
$g = 5$		$g = 5$		$g = 5$	
$\underline{a} = 6$	$\underline{b}$	$\underline{b} = 15$	$\underline{a}$		$\underline{a}, \underline{b}$
$A = 5^6$ mod $23 = 8$		$B = 5^{15}$ mod $23 = 19$			
$B = 19$		$A = 8$		$A = 8$ $B = 19$	
$\underline{s} = 19^6$ mod $23 = 2$		$\underline{s} = 8^{15}$ mod $23 = 2$			$\underline{s}$

5. Now Alice and Bob both have the same key, represented by $\underline{s}$, that is only known to them. Eve, in spite of knowing p, g, A, and B, has no practical way of calculating $\underline{s}$.

The main mathematical foundation underneath this algorithm is the fact that $B^{\underline{a}}$ *mod* $p = A^{\underline{b}}$ *mod p*. Table 4.1 shows a simple example of how this algorithm works for some arbitrary integers. At the end of the cycle, Alice and Bob reach the shared secret ($\underline{s} = 2$) over an insecure channel. Note, however, that in reality the prime number p, as well as secret asymmetric numbers $\underline{a}$ and $\underline{b}$, should be picked much larger to make breaking the code computationally impractical. On the other hand, g is typically small and does not need to be a large number. While deciphering the key is impractical for even large computers, performing the calculations required by Alice and Bob can be done very efficiently.

There are some other use cases and generalizations to the Diffie-Hellman algorithm. For example, it could be generalized to establish a multi-party secret key. Another example is forward secrecy where a new key is generated for any new session between the two parties; that way an instance of compromise does not reveal all the communication history. The latter use case is practically doable because of the efficiency of the algorithm to generate new keys. Over the past few years, some cybersecurity researchers have proposed algorithms that can penetrate this algorithm at budget levels that can be afforded by state-funded agencies in the order of $100 million. This is another reminder that security is a highly relative term. The goal of the algorithms is to make a breach more expensive and time consuming.

Rivest-Shamir-Adleman or RSA Algorithm

One of the most widely used public key encryption algorithms in the industry is RSA. This algorithm was patented in the United States by its three namesakes. It is another one of the asymmetric cryptographic algorithms where the sender and receiver have different keys for encryption and decryption, respectively. Let's start showing the algorithm through a simple encryption and decryption example given the public and private keys. We will then quickly review the process of generating the keys.

Assume that Bob wants to receive messages from other parties, including Alice. He announces the public key to be the pair (5, 14). We will discuss in more detail how those numbers can be generated. Note that everyone including the eavesdropping Eve can hear the public key to be (5, 14). Now Alice decides to send one character B securely to Bob. What she does is to first transform the character to a number (for example based on the ASCII table). Let's for the sake of simplicity, assign the number 2 to our message. Then she uses the key to encrypt the number as $2^5 \ mod \ 14 = 4$. So Alice sends the encrypted message 4 to Bob.

On Bob's side he needs to decrypt the message. His *private* key is the pair (11, 14). Note that the second number, 14, is the same as in the public key but the first number is different and is only known to Bob. Now Bob takes the message received, 4, and decrypts it by doing the reverse process as Alice but using his private key instead: $4^{11} \ mod \ 14 = 2$. So he was able to successfully decrypt Alice's message. Now let's see how the public and private keys are generated.

1. Pick two large prime numbers p and q. Keep these numbers secret.

2. Calculate $n = pq$. n is referred to as the modulus in the key which is common to both public and private keys (number $14 = 2 \times 7$ in our example above).

3. Define $(n) = (p - 1).(q - 1)$.

4. Find an integer e that satisfies $1 < e < (n)$ and that e and (n) are prime with respect to each other (share no prime factors). **e is the first part of the public key (e, n)** which was 5 in our example above.

5. Find an integer, d, such that $de \equiv 1 \ (mod \ (n))$. **$d$ is the first part of the private key (d, n),** which was 11 in our example above.

The underlying reason why this algorithm works is that finding the prime factors of a large prime number is computationally difficult. At the same time encryption and decryption at run time can be done quite efficiently even though large exponents are involved. For applications with a reasonable amount of compute power, a popular choice for the public exponent, e, is $2^{16} + 1 = 65\,537$. However, for small form factors where energy and CPU power is limited, such as many IoT devices, smaller exponents may be chosen for practical reasons. Note that smaller values for the public exponent reduces the hardness of the algorithm. Also most of the security algorithms to date assume silicon-based technologies for processing units. As quantum computing gets closer to reality, many such problems that would take many years for the hackers to solve can become trivially easy. If and when such a revolution happens, security schemas should be fundamentally revised.

While we walked through the algorithm above, curious readers can read through the mathematical proof of how the algorithm actually works. In a practical sense, there are libraries, such as OpenSSL that are optimized to generate public and private keys very efficiently. The main takeaways of this segment is getting a feel for how typical public key algorithms work and what makes them difficult to break. There are many variations of every single algorithm with relative strengths and weaknesses. The main factor to keep in mind is the security hardness of the algorithm balanced by the cost of compute, power consumption, and user experience.

Hash Functions are used to check the integrity of data transmission. They can be used to store passwords without revealing the actual string of characters. A hash function maps a message of arbitrary length to a fixed number of bits that is called the hash value. Approved hash algorithms for generating a condensed representation of a message are specified in two Federal Information Processing Standards: FIPS 180-4, *Secure Hash Standard* and FIPS 202, *SHA-3 Standard: Permutation-Based Hash and Extendable-Output Functions*. FIPS 180-4 includes SHA-1 and the family of SHA-2 hash algorithms. FIPS 202, on the other hand, offers a newer family of permutation-based algorithms that include fixed-length and extendable output functions. A good hash function has the following characteristics:

1. It is deterministic.

2. It is computationally efficient to go from message to hash.

3. It is computationally impractical to go from hash to message easily; sometimes these algorithms are called one-way functions.

4. A small change in the input message generally leads to a significant change to the output message. Therefore, if two hashes look very similar to each other the original strings can be quite different. The hackers cannot infer much from the proximity of hashed messages.

Hash tables for storing passwords are generally considered secure and are widely used. One of the ways for attackers to breach such password systems is when they somehow get access to the hash table. In that case, they can try brute force or smarter approaches to reverse the process and see what string possibly would hash to any one of the entries in the hash table. Once they strike a match, they would be able to use the password string and take control of that user's account.

Message Authentication Code (MAC) is a relatively short addition to the actual message to ensure the authenticity of the sender, as well as the channel. Imagine that Alice and Bob have already established a *shared private key* between themselves. When Alice decides to send the body of the message to Bob, she also passes the whole message through the MAC algorithm using the private key. This process generates a relatively small output called a tag. Alice then transmits the message together with the tag to Bob. When Bob receives the message from Alice he is concerned about two main questions:

- Is someone other than Alice pretending to be Alice as the sender?
- Did the message get manipulated or corrupted on the way to me?

To answer these questions, Bob generates a tag by pushing the body of the message through the MAC algorithm using the same private secret key as Alice. Then he compares this generated tag with what Alice sent over; if the output tag is the same as the tag received from Alice, then Bob can confidently assume that the answers to both questions above are negative. If you are familiar with checksums the idea behind MAC is similar.

SOFTWARE AND FIRMWARE UPDATE TECHNIQUES

Updating the software on IoT devices (as well as many other setups) such as smart sensors, smart actuators, cars, personal wellness trackers, and others is a critical aspect of product design and ongoing software lifecycle management. Embedding such a feature for the car manufacturer Tesla is believed to be one of the significant factors behind the brand's meteoric success. It enables the manufacturer to fix bugs and plug security holes fast over the air at a much cheaper price. Both the users and the manufacturer win. On top of that, it lets the manufacturer release products faster even when certain software aspects are not fully ready. Over time the manufacturer can then release the new features through subsequent software updates. Without a secure and automated infrastructure to push software updates, it is very hard to imagine a scalable IoT system.

Needless to say such software and firmware updates need to happen *securely* after proper *authentication* and *integrity* tests. It is very important for the designers to build the hardware, software, and processes right from the beginning to enable such secure updates. Oftentimes, cash or time-strapped teams cut corners on these down-the-road requirements in the interest of more short-term wins. Changing the platform or process at a later stage, in order to enable a secure and over-the-air software update, gets exponentially more costly and difficult.

Random Key Generation

In the process of software updates, random key generation is critically important for security as we will see below. The keys should be random numbers that vary over time frequently enough. These random numbers will be the basis of security keys as explained in the few algorithms we reviewed. Lazy designs sometimes replace random key generation with pseudorandom numbers that are often offered by programming languages and kernels. Deep down such pseudorandom sequences come from predictable algorithms that will result in compromised security. Pesky and smart hackers can reverse engineer such predictable random generators and find a way to guess them. Good designers try to stay away from such pseudorandom generators.

There are a few ways to generate random numbers in real systems. In the case of embedded systems, ring oscillators are sometimes intentionally built into chips to offer random sequences. Imagine a chain of inverters where the output of the i-th one becomes the input of the $(i+1)$-st inverter. The very last one is then connected to the very first one to make a ring. These inverters are intentionally placed in a scattered way across the chip so that the signals are prone to random noise from other signals in the environment; therefore, they can flip along the way. The sequence of the outputs of these inverters across the rings is a reasonable true random generator on a chip. Other ideas for random generation include measuring certain environmental and physical quantities such as temperature, vibration, pressure, or flow when such measurements exist and then use them as the seed for a random number.

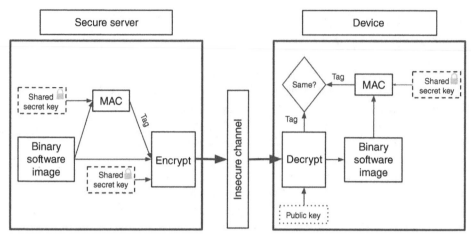

Figure 4.7 Software update architecture.

Software and Firmware Update Process

There are several ways to build a secure software and firmware update system. Below we review one such approach that offers authenticity (who), integrity (quality), and security (safety). The approach leverages an asymmetric public–private key pair. It also uses Message Authentication Code (MAC) on the binary software package. The process has two parties: the server that offers the software update and the remote device that receives the update. You can follow the process along in Figure 4.7 for better clarity. Initially, the two sides establish the public and private key pair among them using one of the approaches we discussed earlier. The server also generates MAC out of the software binary package leveraging the private key. It then encrypts the software package and the tag together using the private key. The encrypted message is transmitted to the device.

Upon receiving the encrypted package, the device first unpacks the message (including the tag) using the corresponding public key. It then uses the private key to generate the MAC independently; if the resulting MAC matches with what was received from the sender, the package is accepted. Otherwise, the process is restarted.

KEY MANAGEMENT AND PROTECTION AGAINST KNOWN THREATS

By now we have established the notion why security is a must and how most of the existing algorithms rely on public and private keys to communicate with other machines over the insecure network. Eventually, what the algorithms do is making it more difficult for the adversaries to get in our systems. In more technical terms, the security algorithms' role is to harden the threat surface by making it computationally very expensive to decipher the ciphertext into plaintext. At the risk of sounding like a broken record, we need to always think like a hacker: in any design decision, try to step out of your normal way of thinking and get creative. Think how the design can be broken or tricked.

Mathematically speaking each algorithm makes it extremely unlikely for a random guess to be able to decrypt the encrypted data. One of the great examples of this is the AES block encryption algorithm family. It is commonly used to encrypt stored data on disk or flash memory. The key can be a 256-bit string that is produced by a random generator on the device as we discussed earlier. For a hacker to get lucky and guess the key through brute force it means he needs to guess the right string out of $2^{256} = 1.2 \times 10^{77}$ possible combinations. Using a million computers to try one code every nanosecond, it would take 10^{54} years to try all possible combinations.

Now continuing to think like a hacker, we should ask ourselves if brute force doesn't work, what other ways might be out there to exploit? If guessing the key is too hard, the hacker may attempt to steal the key. This brings us to the notion of *key management*. Similar to our personal passwords, there are a number of general best practices. Each of the following factors increases the vulnerability of the key:

- The more places a key is used (as opposed to only in one device or system)
- The more places it gets stored
- The longer duration we keep the keys unchanged

Key management and the foundational encryption algorithm should be treated quite differently. The former is both an art and science; any organization can construct a secure process that fits their needs and security requirements as to how to manage their keys. However, we certainly do not recommend getting creative with known and standard algorithms to encrypt the data itself. These algorithms have been developed with lots of care and study behind them, so it is more likely than not that manipulating them would create holes in them. Another very important reason is that customers and users increasingly expect security certifications by known standard bodies and security organizations such as NIST. Those certifications typically follow the steps of defined and published algorithms quite strictly. Deviating from the algorithms may cost a product or piece of code dearly by causing it to disqualify from a certification.

One of the common techniques to protect the keys is called *wrapping*. It means the key owner would wrap her keys in other safe passwords and save them in a safe place. A simple implementation of this is to encrypt each key with one highly secure and continuously changing password and then only save the encrypted versions in a safe database, which is also behind a password. Even if the database's security is compromised, the individual keys are still one decryption removed from being exposed.

SECURITY ATTACKS AND CHAIN OF TRUST

Technology is only one aspect of making an IoT system secure. In what follows, we look at a few different practical concepts that span technology, people, and processes in the context of security.

Security Attacks

So far we defined and reviewed several security concepts and reviewed their pros and cons. Now let's turn our attention to a few forms of security attacks.

Man in the Middle Attack: in this form of attack, Bob and Alice are to communicate with each other through a channel. Bob uses a typical duo of public and private keys. Imagine that Bob publishes his public key. Bob uses his own corresponding private key and sends an encrypted message. If Eve, the attacker, can disrupt the physical medium and get a hold of the message she becomes the (wo)man in the middle. She can then use the public key to decrypt the message, insert her own desired messages or just record the communication, and then encrypt the message again with her own private key and resubmit to Alice who is the original intended receiver. In this form of attack, the two communicators may not notice the existence of the man in the middle for a while which makes this attack difficult to defend against. Hardening the communication channel in the middle where the bits transit is a general effective tool in this case. Another potential way to do this is to make the "public key" less of a publicly known key by establishing it between Bob and Alice through a special channel that is known to be highly secure and inaccessible to Eve.

Another way to tackle this issue is leveraging the Security Certificates. This model is very common in SSL and HTTPS communication in most web browsers. A central trusted authority establishes itself as the source of trust. In the offline initial step, the Certificate Authority (CA) receives a request from servers to issue a certificate. The CA then encapsulates the basic information about the server and its identity, and encrypts it with its own key. The CA then publishes the certificate to the web browsers usually in the form of hash tables. Now when the server goes online client nodes start an SSL session with the server; to establish the trust for the session the server sends an encrypted certificate to the client. The client decrypts it with the server's public key and then compares it with what it has already received from the CA in its hash table. If they match, the server's authenticity is established. Then the client shares a random key for that session for the rest of the communication in that session. While quite effective, the cost of this approach is reliance on a central authority.

Replay Attack: imagine that an authorized person or machine is sending a command to a destination machine. It can be to shut a valve or open a gate or turn a machine off. The receiver needs to know the authenticity of the sender. Therefore, it uses a Message Authentication Code (MAC). So far things are good and a random hacker cannot replicate the secure command. However, a listening agent can eavesdrop on the whole message, which includes the command and the encrypted MAC, and record it for future use. At a later time, it can pretend to be the original sender by replaying the same stream of bits and get the command executed at will. To defend against this attack, one can essentially make MAC a time-varying encryption mechanism. Using a concept called Nonce (short for Number Used Once), the legitimate parties can agree on generating a series of numbers at each instant of time and inject that into their MAC encryption and decryption processes. The less predictable the Nonce, the more secure the communication becomes.

That way even if the eavesdropper replays the exact string of bits from yesterday, the MAC check would fail at the receiving side because today's Nonce is different from yesterday.

Memory Protection: one of the best practices in building any application is memory protection. This principle applies to security, as well as protection against application bugs such as memory leaks. In short, memory management allows an application to access only a subset of the whole memory available. Leveraging the Memory Protection Unit (MPU) that is offered by certain CPUs keeps malware and injected code from claiming the whole memory or reading data that belongs to other applications. Let's see how one major force in the world of chip design has enabled memory protection. Arm-based chips have been used in 100 billion devices as of 2021. The chip company estimates that the number will rise to 300 billion over the following decade. The MPU on ARMv8-M processors supports up to 16 regions. The feature limits the potential damage caused by application bugs. The company also announced the concept of Realms in early 2021 as part of the Armv9 announcement; these memory Realms, which are an aspect of implementing confidential computing, shield a certain part of the memory from the operating system, as well as other applications. Therefore, developers can build applications whose data and code is protected and also are immune from unintentional mistakes that can harm other applications.

It is also highly recommended to manage application permissions wisely. Applications typically only need access to system resources at a *user* level; only system applications such as firmware or kernel code requires access at an *administrative* level. Being too generous when it comes to granting system permissions can end up being costly over time. Sometimes lazy designs that tend to avoid the nuances of a good security structure grant too much freedom to all applications. While it can make the short-term life of the developers easier, over time it can result in many different types of security and quality issues.

Two very famous vulnerabilities affecting several Intel and Arm-based processors, among others, were discovered by a number of benign actors. However, whether or not they have been used secretly by malicious actors is not known. Table 4.2 briefly introduces Meltdown and Spectre vulnerabilities.

In order to protect against these vulnerabilities, we recommend a few protective measures.

Physical Protection: in certain cases, simple low-tech protective and tamper-proof stickers or conformal coating can be effective. Such measures are used to show if a piece of equipment has been tampered with. However, beware of the hidden cost of some of these physical protective measures; troubleshooting a device that is covered in epoxy resin as a protective measure can be extremely difficult.

Software Protection: there are a number of best practices that always need to be applied.

- All communication channels need hardening. The hardening ranges from physical protection to software protection such as encryption, authentication, and authorization as discussed earlier.

Table 4.2 Meltdown and Spectre vulnerabilities explained.

Meltdown breaks the barriers that are supposed to keep one application's memory isolated from another one's and hence, allows attackers to steal secrets such as passwords, personal data, and other critical information.

Spectre tricks normal applications to consider as their memory a part of system memory that is not actually theirs. That leads to these applications writing their data in areas that can be read and abused by a malicious application.

- Always check the port number, IP addresses, or other logical addresses of messages to make sure they make sense and are in an acceptable range. Never assume that the senders of the messages are necessarily who they claim they are.

- Security protocols and design reviews with security in mind should be applied to all design, development, and production stages of hardware and software products.

- Unexpected channels are sometimes the most dangerous threat vectors. Such channels include power and RF intrusions where the hackers try creative ways to turn power and electromagnetic readings into smart guesses. Even rapid changes in ambient temperature can disrupt normal operation of chips leading to malfunctions. Some malfunctions lead to security measures to break and pave the way for the attackers to gain access to data. There are a whole class of attacks that use RF analysis of signals over the air, power glitches, and other ways to read data or put the system in a compromised state. These are categorically called Side Channel Attacks. The main lesson is that attackers think differently. You need to think like an attacker to reduce the likelihood of security compromises.

Chain of Trust

One of the significant costs of applying security is loss of flexibility. Taking any action like opening a new port for interaction, allowing new users to log in, or running code opens up new threat vectors. One of the ways to mitigate the erosion of flexibility is the concept of Chain of Trust.

Let's use an analogy. You are likely to believe your friend that you trust when she tells you that she has read in the *Wall Street Journal* quoting the CEO of a major company about their upcoming product. You may even take action and buy or sell their shares based on the news. Now if you rerun the story, the immutable source of truth about the company is its CEO, so we are extremely unlikely to be making a mistake there. The next in the chain is the *Wall Street Journal*; we tend to trust the *Journal*. But note that, when it comes to this specific piece of news, they get their credibility from the link up, meaning the CEO. And then your friend is a trusted source for you. And similarly, she is trusted with the news about the company, due to the fact that she is quoting the *Wall Street Journal*. Hence, your deduction is reasonable and likely true. If you decide to become an idealist and seek direct confirmation by the CEO, you won't get anywhere; no reasonable person expects to track and confirm every single piece of information all the way up to its original source. It would be highly impractical.

Now let's get back to our security concepts. In the life of an IoT device, since the moment it's turned on, several different actions take place: a piece of software called the bootloader which is usually etched in the hardware, moves itself to the working memory. It may then call for the bulkier and more potent pieces of software at the boot time from other sources such as external memory or disk. When the initial boot is done, other higher-level applications may start running in an order to bring about different functionalities. And finally during the normal span of the device's running time, new applications and users may interact with the device, fetch information, execute commands, or inject data. All of these steps need to happen in a secure and authorized manner. At every step of the way, the higher link in the Chain of Trust needs to authenticate the following link per some of the standards and methods we counted earlier (such as MAC). The bootloader on the device is generally treated as an immutable and trusted source of trust; certain private keys may be loaded in the Read Only Memory of the device. The external bootloader then has to be authenticated, and subsequently authorized accordingly, before it can proceed. And the applications in turn need to be authenticated by the bootloader; and the chain continues. Note that it would be utterly impractical to expect the high-level user or application to go all the way down to the lowest level of hardware for authentication. It would bog down several aspects of an efficient piece of IoT device.

Incident Response

Even after applying all precautionary security steps, breaches and incidents are all but inevitable. The public news and security forums are filled with small and large such incidents. So, while organizations and consumers are advised to do all they can to prevent such events, preparing to respond to events of harmful nature is the other must. Such harmful incidents usually span both cybersecurity breaches, as well as discovered software bugs. The United States National Institute of Standards and Technology (NIST) has a very well-written playbook titled "Computer Security Incident Handling Guide." We recommend anyone who is seriously considering offering IoT products or services to read and practice this playbook. It breaks down the response best practices into three layers of policy, plan, and procedures.

The response generally is broken down into two major pieces. One piece deals with the handling of the very issue at hand. It starts from the discovery of incidents. What is the scope of such incidents and who in the organization has the responsibility and authority to make such determination? Once that determination is done how should they be classified? How many levels of urgency make sense for the organization? Are there potential legal ramifications where the legal representation is required? Who has the responsibility to actually research the technical issue at hand and make recommendations for fixing? These are all important questions that the first aspect of the response plan needs to address.

Equally important is the messaging and communication plans around those incidents. When data is breached or thought to be breached, or a new security hole is discovered, many parties may be interested. As is the case for most kinds of crises, reliable, consistent, and trustworthy communication is extremely important. Starting from the customers, the organization needs to have a solid and already-prepared plan to communicate the right and crystal clear message to the customers. If the product is incorporated in other partners' products or services, the messaging to those partners is also very important so they can move quickly and inform their customers. In many cases, government and law enforcement agencies need to be brought in the loop for compliance reasons. In those cases, a well-crafted legal language wrapping the technical details is advised. Last but not least are the employees of the company. In the absence of trustworthy and commanding communication from a central response team, the rumor mills can start mixing truth and myths together leading to negative morale or public relations disasters.

As you can imagine, being prepared to get all of the steps right in a professional and poised manner is no small task. If not tackled properly, the reaction might do more damage than the actual incident. Most serious organizations have dedicated teams in place to handle incident responses. Clearly defined roles and responsibilities are in place so the chain of command at the time of crisis is already established. Oftentimes the RACI model is used to predetermine who is Responsible, Accountable, should be Consulted, and should be Informed in such incidents. Savvy organizations use external resources, including in most cases governmental assistance, to educate their core staff. These methodical teams start from the detection and analysis of the incident, and move on to containment and eradication. And finally post-incident gear kicks in where the long-term lessons are crafted and put in practice. None of these steps is trivial. In fact, in a real-world story in 2021, a victim company ended up paying hackers a hefty amount of ransomware twice in a matter of weeks just because they forgot to close the loophole post-event the first time. Don't be that kind of company!

BLOCKCHAIN

Blockchain is in essence a distributed consensus algorithm. Although the general idea was around for a few decades, the main formulation as we know it was proposed by an individual or entity nicknamed Satoshi Nakamoto in 2008. His or her or their true identity is still unknown as of writing these words. The concept leverages consensus around facts that get recorded in many copies of an electronic ledger. Each fact or transaction including its description, parties involved, and timestamp makes a block. Every node in the distributed network of blockchain records the transaction by adding it to their ledger. What makes this proposal unique is that a security hash of the previous block is also included in the next block. In other words, each block of facts is dependent on the block behind it, and by extension, all the blocks before that. This chain of blocks makes the blockchain. Therefore, tampering with any block after the fact makes everything after it invalid and detectable. All the nodes in the network keep a copy of the ledger. When the majority of the nodes reach the consensus that a block is valid, it is known to the network as valid.

The most famous use case of blockchain is the cryptocurrency Bitcoin. Since its inception its valuation has grown exponentially. The main ideas behind a currency, such as a dollar bill, are two main factors: general acceptance of the token (dollar bill in this case) as a representation of value and difficulty in reproducing it (hence avoiding double expenditure). Bitcoin solved the second problem by making the *mining* of any new bitcoin tied to solving a difficult security puzzle and adding the network consensus to the algorithm. This is also the main idea behind the so-called non-fungible tokens (or NFTs) where a unique piece of art or the first gif or the very first tweet by an individual is uniquely owned and traded. As for the former factor, the general acceptance of the cryptocurrencies has been on the rise, leading to an exponential value increase. In 2021 for the first time, several major companies, such as Tesla and some big financial names, as well as the country of El Salvador, announced plans to officially accept or invest in the currency.

While Bitcoin is the popular face of blockchain, in reality the use cases go way beyond cryptocurrencies. This is especially true for the industries where there is a long supply chain including multiple entities and geographically distributed sites. Imagine a restaurant that wants to be sure of the quality and freshness of the tomatoes it serves to its customers. Traditionally, threading all the hands between the restaurant and the farm has been virtually impossible given the complex nature of the supply chain. Blockchain, especially optimized for such use cases, provides a way for the farmer, movers, wholesalers, distributors, and restaurants to share one ledger among themselves safely and securely.

In such a ledger, each event from the farm onward gets added to a blockchain that is immutable and transparent to all parties involved. That way each entity in this supply chain has access to true and trusted information at all times.

A blockchain can be public or private. In the public type anyone can join the chain, see the ledger, and become a node. The identity of each member can be revealed or it can be done anonymously. Bitcoin is one such example. In a private blockchain only a limited group of entities can participate and access the ledger. Industrial blockchains are usually of this type to keep the data behind security and privacy walls. There are commercial offerings of industrial blockchains by companies like IBM. Typical industries who can benefit from this technology are food and beverage, healthcare and pharmaceuticals, mining and precious metals. What is common among these industries is that trust in how a certain action has (or has not) happened and under what conditions is extremely important. Also, they tend to be highly regulated industries where providing accurate information about the supply chain is of utmost importance.

While blockchains add a significant level of security and trust to the flow of information, they do not replace the best practice cybersecurity techniques at all. In fact, some blockchains have been the targets of successful attacks in the past. Therefore, all principles mentioned in this chapter such as key management, secure updates to the code base, hardened communication channels, authentication and authorization, incident planning and management, as well as all other best practices are all required in blockchains. Below are a few examples of how a blockchain can be attacked.

- **Phishing:** One of the typical methods of attack is phishing. In these attacks, the victims are targeted by emails that look like sent by the right authority that handles the blockchain. The recipient enters their credentials after clicking on seemingly benign links not knowing that they are in fact giving away their credentials to bad actors on phony websites.

- **Routing:** it is very important for the traffic in a blockchain network to travel through secured and trusted links. In a routing attack, the bad actors penetrate the networking infrastructure and reroute the traffic so that they can eavesdrop on the communication.

- **Sybil:** in the so-called Sybil attacks, the hackers flood the network with requests and traffic bringing it to a halt or crash. In many such scenarios, the security walls also don't function properly leading to leakage of private data.

- **51% attack:** the consensus, or 51%, is the litmus test for a block of information to be considered fact in a blockchain. Some hackers add or hack enough nodes so that they gain control over more than half. It gives them control over executing what they desire.

While the technology is appealing in many industrial and financial scenarios, it does not come for free. The service itself is a specialization that would require hiring or outsourcing at a cost. Certain blockchains are also notorious for their hunger for energy. All those consensus-building puzzle-solving activities are expensive operations when it comes to power. In a February 2021 research, Cambridge University researchers calculated that

Bitcoin uses more energy than Argentina

If Bitcoin was a country, it would be in the top 30 energy users worldwide

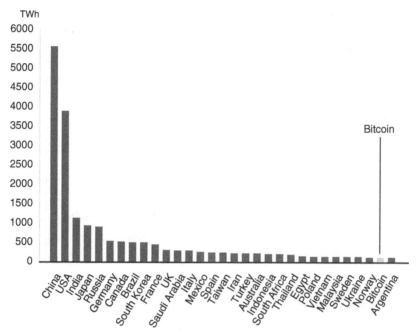

Figure 4.8 Ranking of energy consumption of top countries relative to global Bitcoin operations.

Bitcoin consumes more than 121 Terawatt-hours globally. If global Bitcoin operations were a country it would have ranked thirtieth on the list of countries sorted by power consumption right above the whole country of Argentina. Figure 4.8 shows the bar chart of energy consumption by the top 30 countries plus Bitcoin operations. Therefore, like any other business or technical decision, the special need to adopt this technology should be studied and the costs and benefits should be evaluated.

SUMMARY OF BEST SECURITY PRACTICES FOR IoT SYSTEMS

So far we introduced several security concepts and made recommendations to harden IoT systems. Before we turn our attention to some other aspects of networking design, let's summarize those concepts all in one place.

- Develop a security mindset that invites different ways of thinking. The designers and developers only create security along with their thought process. New eyes bring new ways that security can be broken.

- Always clearly define what is to be protected:
  - For information and communication use encryption algorithms.
  - For authentication use MACs to make sure different sides are who they say they are.
  - To protect the integrity of sensitive information use hashes.
- Always use the current algorithms that are considered the most secure, tested, and accepted algorithms at the time. Resist the temptation to create brand new but untested algorithms that seem smart and secure.
- Security should not be an afterthought or the responsibility of only certain people. Everyone in the chain of production should play their roles.
- Key management is an essential part of building a secure system.
- Providing software updates is both important and nontrivial. While it matters a lot to keep software and firmware up to date, it is also essential to build a secure process to make sure the right package gets authenticated and installed.
- There is no perfect security. As systems and software change, there are always new ways that systems can be penetrated. Creative, patient, and resourceful hackers are constantly innovating new ways to break into systems.
- Form incident response teams whose job is to define processes for post-incident actions and communication for all interested parties.
- Check NIST and FIPS Computer Security Resource Center for tested and published best security practices.

NETWORKING CONCEPTS

In this segment, we continue our journey in security to cover some of the important networking concepts in an IoT system. Most of these concepts are more relevant to industrial and business IoT environments. However, being familiar with the technologies is of value for any IoT professional because of their performance and security ramifications. As we will see under each technology, each can offer some level of security at the network layer and harden the system further. These concepts are complementary to the encryption and key management concepts that we discussed earlier.

Proxy

A proxy server, as the name suggests, acts as a proxy between the nodes in a network, machines or humans, and the global Internet. When an IoT device wants to send data to a remote server or request a service, it has to communicate with the server. In a proxy set up, such communication first goes through the proxy server (Figure 4.9). The proxy server then passes on the communication on behalf of the IoT device. When the response comes back from the remote server, the proxy server receives the message and passes it back to

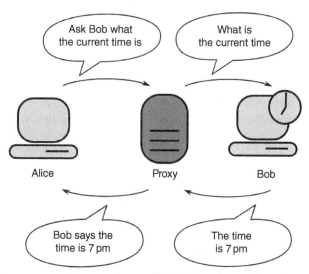

Figure 4.9 Proxy server (*Source:* https://en.wikipedia.org/wiki/Proxy_server#).

the actual recipient (the IoT device). There are a number of use cases for proxy servers especially for security and privacy purposes.

The proxy server can mask the IP address of the actual sender. That way the identity and IP address of the sender is not revealed, hence the privacy is boosted. In some larger organizations or even small households with minors, a proxy server can enable a level of control to the network administrators and parents. Since all the traffic has to go through the proxy server, the proxy designer can apply certain rules and block certain types of content. Good proxy servers also block certain incoming traffic when they detect security risk due to the sender's identity or upon packet inspection.

In certain countries, some web resources are blocked for the population. A proxy server can circumvent such restrictions: the individual user masks its request for the forbidden server in a way that it seems to be happening only between the user and the proxy server. Good proxy servers also encrypt the communication over the Internet. That way, both the security and privacy of end users are improved. When it comes to the level of anonymity, proxy servers offer various options. Some are transparent: they only act as the proxy and reveal the actual IP addresses of the individual nodes to the recipient. In another mode, the proxy server declares that it is a proxy and that it is hiding the IP address of the requestor upon its request. Distorting proxy servers declare that they are proxy servers and reveal a distorted version of the user's IP address.

Another benefit of proxy servers is related to the bandwidth saving and performance boost in the network. If a website or data set is commonly requested by the nodes in the network from a remote resource, a cached version is kept by the proxy server. That way the roundtrip to the actual remote server is saved resulting in both bandwidth saving and latency reduction.

Having said all of the above, we should add that not all proxy servers are created equally. Some inexpensive public proxy servers may pose privacy risks of their own by exposing users to data breaches. If they are not offered by trusted providers, by virtue of having visibility into all the traffic, they can actually pose major privacy and security risks to the users. As a rule of thumb, there is not much justification for a proxy that does not offer encrypted communication. Also, if the computing and bandwidth resources are not optimized, they can become the bottleneck in the communication to the Internet and cause latency.

VPN

A close relative of the proxy server is virtual private network or VPN. You can think of VPN as a secure tunnel between a node and a server, or between two sites. When a node connects to a server through a VPN connection, the communication between the two is encrypted and hidden from the ISP. All the data gets encrypted in a way that the IP address and the content of the communication is private and safe (Figure 4.10). A common use case for VPNs is when employees work remotely far from their office. In that case, they connect to their office network through what is called an SSL VPN. In this scenario, an employee's remote device gets securely tunneled to the server at the company site as if it is on the company network. This is extremely important for work that involves business-sensitive material; if a hacker can sniff the traffic between the home and office, they will not be able to decrypt the content easily.

Another type of VPN is site-to-site VPN. In this scenario, a business can have multiple sites that are physically far from each other and hence own their own local network. For the employees that work at these sites to have seamless access to each other's content, a site-to-site VPN can be used. This type of VPN tunnels two local networks securely, so a node on one can be treated as if it is also on the other network.

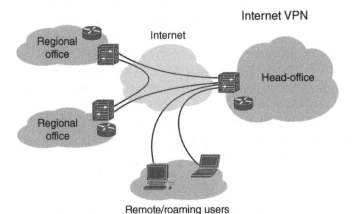

Figure 4.10 VPN (*Source:* https://en.wikipedia.org/wiki/Virtual_private_network#/).

A VPN not only encrypts the data it also masks the IP address of the sender. The VPN server communicates the messages to and from the Internet and then communicates the packets with individual nodes connected with it in an encrypted fashion. Therefore, neither the hackers nor the ISPs or the government will know easily what the node is communicating in and out. A good VPN also provides features such as "kill switch" where upon detecting a malware or threat, it automatically closes certain sensitive applications on the client node to minimize the risk of exposure.

Sometimes, VPNs are used to help resolve some geographical limitations or provide compliance. When asking for certain data sets from a certain geography is not allowed the client node can connect to a VPN server within the allowed geography, make the request through the VPN as the proxy, and then consume the data securely through the encrypted tunnel.

While there are some similarities between proxy servers and VPNs in reality they are not the same. A pure proxy server only acts as a proxy although some offer additional privacy and security features as well. A VPN, on the other hand, has encryption features at its core. The level of encryption also differs between the two. A VPN encrypts data at a lower operating system level while a proxy typically works on an application by application basis and only reroutes traffic for an application at a time. For these reasons, typically VPNs are considered safer. The cost is that they are often slower due to heavier encryption operations. They also generally cost more than the proxy servers.

Firewall

A firewall is a technology that is put between a secure and an insecure network for protection. The term originates from the metal pieces that are used between two sides of a building to shield one side against fire that starts on the other side. It was applied to network security concepts starting in the 1980s. While a local area network (LAN) of a house or wide area network (WAN) of an organization is considered safe and friendly, the wider Internet is filled with attacks and insecurities. Hence, there is a need for such a wall of security between the two sides of the network. Figure 4.11 shows a logical view of how a firewall sits in between the two sides.

A firewall can be set up to reject traffic by default unless certain rules are met. This approach is called whitelisting; a certain type of traffic should be explicitly whitelisted before it can pass through. Alternatively, a firewall can let the traffic through unless a certain type of traffic is blacklisted. The action taken by the firewall can also be configured in specific terms. It can simply throw away the packets silently, return the rejected packets to the sender, or route the packets to a pre-specified set destination for inspection or further processing. Physically, firewalls come in various shapes. Some are deployed on especially-made pieces of hardware dedicated to the firewall functionality. In some other cases, they are pure software products that can be installed on servers or virtual machines. Architecturally firewalls can be put in front of the whole network and apply blanket rules to the traffic associated with all of the nodes inside. This is quite common for business environments. Other forms of the application are deployed on individual devices. Windows

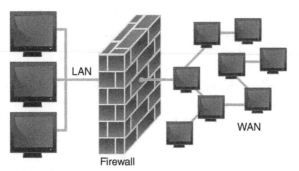

Figure 4.11 Firewall (*Source:* https://en.wikipedia.org/wiki/Firewall_
(computing)#).

10, for example, comes with its own embedded firewall. Such software firewalls often run
as background processes.

The granularity of the blockage rules can vary quite a bit from one technology to
another. Modern firewalls allow the rules to be set based on source, destination, port
number, or application. Some of the newer technologies allow for deep packet inspection to
identify a wider range of threats to the networks.

Note that like most other security aspects, a firewall adds another layer that the packets
need to go through which means extra latency. There is no doubt that in most cases the
benefits outweigh the risks; however, it is imperative that the network engineers on the
one hand and the IoT designers on the other hand are aware of such trade-offs. There are
some mitigating techniques that can be applied to minimize the latency. For example,
certain ports, which are very easy to check, can be used in the devices and then opened up
in the firewall, so that the traffic can pass through with very little added delay.

DMZ

A demilitarized zone or DMZ is a zone between two neighboring but not necessarily
friendly sovereignties where military activities are forbidden. The main purpose of a DMZ
is to provide further security between the two sides by creating a buffer. For a resident of
one of the sovereignties, the DMZ is not as safe as inside his territory but still safer than
the other unfriendly territory.

The analogy is used in enterprise networking concepts and for many IoT applications
because there are internal and external networks: the inside of an organization often
has a LAN or enterprise WAN where everything is tightly in control, sensitive servers
and databases are kept, and the accessing people are trusted. On the other hand, the
broader Internet is filled with threats. You cannot completely isolate these two domains
from each other in the same way that you cannot completely isolate the two neighboring
sovereignties. On the one hand certain internal services may need access to external
services such as weather forecast services, bank servers, or other service providers.

Employees will also need to be able to access the public Internet for their personal affairs from time to time or to take a break. In the reverse direction, external services or humans, such as partner servers or individual customers, may need access to the corporate data and services.

A DMZ is one of the architectural ways to enable data exchange and services between inside and outside of an enterprise while providing a significant level of security. While internal sensitive databases and industrial analytical servers need to be protected from any external access, the corporate's mail server, FTP server, and web server all need to be exposed to the external Internet. Therefore, the latter group is placed in the DMZ. One side of the DMZ interfaces the external Internet, thereby allowing access to web and email servers to the customers. On the other side, it interfaces the internal secure network. Note that certain customer-facing services need to fetch data from sensitive operational databases or IoT devices to show the customers their data or provide a servicing partner with what they need to operate. That way, the DMZ is less secure than the internal LAN but more secure than the open Internet. The main technology used to implement a DMZ is a firewall.

There are two common architectures to make a DMZ. One is less expensive, less complicated, and less secure. In this single firewall architecture (Figure 4.12), one firewall acts as a hub while logically routes, blocks, and allows traffic between the three connecting networks through its network interfaces: outside, DMZ, internal. That way all the traffic goes through this one point of failure.

In this architecture, while things are kept simpler, the one firewall becomes one single source of failure. In the more common and more secure architecture, two firewalls are used (Figure 4.13). In this double firewall architecture, one firewall is placed between the outside world and the DMZ and another one between the DMZ and the internal network. One of the benefits of this architecture is better modularization of the network. Also in the event that any of the firewalls fails or is compromised, there is still another layer of defense between the outside world and the internal network; therefore, this architecture offers more security. In some cases, organizations opt to go for firewalls from two different vendors. That way it takes two vendors to miss a threat together for it to become a security hole.

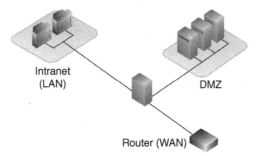

Figure 4.12 Single firewall DMZ (*Source:* https://en.wikipedia.org/wiki/DMZ_ (computing)#/).

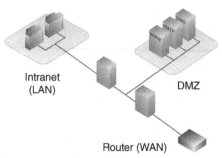

Intranet
(LAN)

DMZ

Router (WAN)

Figure 4.13 Double firewall DMZ (*Source:* https://en.wikipedia.org/wiki/
DMZ_(computing)#/).

Some organizations also include a two-way proxy server inside the DMZ. The benefit of having such a proxy server is to make sure all internal traffic goes through the DMZ before it hits the insecure outside world. It also makes sure that the externally originated requests all go through the DMZ before reaching sensitive servers inside the enterprise. These proxy servers are often combined with an application-layer firewall to allow or deny access very surgically. For example, instead of allowing all the traffic that meets certain IP address requirements, they can allow only traffic from a mail server in the DMZ reaching a database inside the enterprise to craft monthly updates to customers.

Certain standard organizations or leading digital companies provide recommended architecture designs for IoT networks. In these reference architecture proposals, various aspects of an IoT network are designed, tested, and proposed in a way that provides scalability, performance, as well as security. For example, Cisco publishes several such architectural references for networking needs of industrial IoT scenarios. Using these reference architectures can help save time, make networks more secure, and also future-proof designs against nonstop changes in the technology to a large extent.

IoT SECURITY STANDARDS AND CERTIFICATES

When consumers or business users decide to purchase devices they often rely on standards and protocols that they trust. That way, without having to be experts in security, they trust the stamp of approval from an unbiased expert that is trusted to gauge the level of security in a product. This is in particular important in the B2C market where a vast majority of the consumers will not have the expertise or resources to properly assess the security of an IoT device. At the same time, B2B users are also highly encouraged to pay attention to these widely accepted standards before choosing an IoT technology. In the concluding segment of the chapter we review some of these standards and certification programs.

ioXT

One such standard is called the Internet of Secure Things or ioXt. There are tens of large and small member companies in this standard alliance such as Google, Amazon,

Facebook, Comcast, ARM, and many more. Its mission is to build confidence in Internet of Things products through multi-stakeholder, international, harmonized, and standardized security and privacy requirements, product compliance programs, and public transparency of those requirements and programs. The main focus of the standard is on the manufacturing and maintenance processes as opposed to suggesting protocols.

The alliance has produced a public pledge document. In the pledge, there are three main categories in which the takers commit to certain principles. Below we briefly review these principles. We encourage the interested user to visit the ioXt public website for more details and the text of the pledge.

Security

- No universal passwords: the standard specifies that common and universal passwords such as "12345" or other typically easy-to-guess passwords for the user and administrator accounts should be avoided. It is recommended that users are required to change the default password at the setup or first time use. Also, for products that have a user-facing interface two-factor authentication is advised.

- Secured interfaces: all the interfaces to the device, physical and programmatic among others, should be protected against attacks without any assumption of security. The attack domains include remote (programmatic through network), proximity (based on RF or when an attacker takes control of the user's device or network), and physical (when the physical device is in the attacker's control). In all such instances the pledge takers will take specific steps to make it increasingly difficult for the attackers to access the user's data, control its operations, or leverage the device to attack other aspects of the user's network.

- Proven cryptography: as we discussed in more detail, security keys are paramount to the security of an IoT system. Various algorithms exist, some of which are tested and approved by independent trusted bodies. This item of the pledge requires manufacturers to use the known protocols to make it increasingly hard for attackers to compromise the IoT systems. It is encouraged that the implemented algorithms get tested and verified by independent bodies.

- Security by default: the default mode of operation should be secure. That is in contrast with when the user is responsible for taking certain steps to make the device secure. An example of this is when a phone defaults to password-locked screen.

- Signed software updates: the manufacturer needs to have a thorough and secure update policy to its software patching process. Each software package needs to be encrypted and signed properly per the standards we introduced earlier in this chapter. Rollback of software versions is discouraged.

Upgradability

- Automatic security updates: upgrading the software of all connected devices is a must. All such devices should receive secure software packages without the user asking for them. In this managed scenario, the updates need to be applied automatically. If the user needs to maintain control, they should be notified clearly that they are in control.

- Verified software: the manufacturer needs to have a thorough and secure update policy to its software patching process. Each software package needs to be encrypted and signed properly per the standards we introduced earlier in this chapter. Rollback of software versions is discouraged.

Transparency

- Security expiration date: all devices need to clearly show how long they will be supported for upgrade and security purposes. As new vulnerabilities appear all the time, the user needs to be explicitly notified when their device falls out of such security updates. End of security life is a very important aspect of maintaining a secure IoT system.
- Vulnerability reporting program: the participants are encouraged to have a public reporting program where anyone can report security issues in the products. Bug bounty programs are also encouraged, so the security professionals and the public are encouraged to seek such faults. Having a transparent policy to report discovered issues is also part of this aspect of the pledge.

UL

Another security standard for the IoT industry is UL. UL is a global nonprofit organization based out of Northbrook, Illinois. Their mission is twofold: first to help IoT manufacturers and designers to build more secure products. And second, to relieve users and consumers of the heavy burden of analyzing their purchases for security on their own. The idea behind the company was conceived in 1893 by William Henry Merrill, Jr., who at the time was a young graduate of MIT in electrical engineering. When the World Fair was held in Chicago in that year, he proposed his idea to test electrical circuits and assess them for safety. He managed to get funding from the insurance industry, and that's how UL was officially born in 1894.

An IoT product can get UL-certified by going through a certain security assessment process by UL. In order to give the consumers a range of options, they offer various levels of certificate that is nonbinary. UL certificates cover several industries such as automotive, buildings, chemical, energy and utility, financial services, healthcare, industrial products, retail, and technology.

Matter, Formerly Known as Project Home over IP (CHIP)

We introduced the standard in more detail earlier in the book. In this segment, we briefly remind ourselves of the main pillars and turn our attention to the security aspect of it. Project Home over IP or CHIP was kickstarted in December 2019 by a few big names in the smart home industry including Amazon, Apple, Google, and Zigbee Alliance at the time. In 2021, CHIP rebranded itself to Matter while the Zigbee Alliance called itself Connectivity Standards Alliance. The main motivation was to bring interoperability to a world where no two devices in the smart home necessarily worked with each other. That problem made consumers' experience hard because it locked them in one vendor's ecosystem, for example Google Assistant or Amazon Alexa or Apple HomeKit. It also made the life of the developers and manufacturers difficult because they needed to build different versions of their products for the likes of Google Assistant, Apple HomeKit, or Amazon Alexa. On top of all that there was a lack of security standards and best practices in the industry. We will briefly review the status of Matter as of the year 2021. Matter's focus is on much more than just security; we will review its high-level concepts before focusing on security. Interoperability, required best practices for security, as well as provisioning of a ledger based on blockchain for lifecycle management, are all aspects that make Matter-certified devices more secure in general.

Matter is a set of application layer open source protocols for interconnection of appliances (Figure 4.14). The Internet Protocol (IP) is chosen because it has been a tried and tested protocol for many years upon which TCP and UDP have been developed for the transport layer. The networking technologies that are initially supported by Matter are as follows:

- Wi-Fi for high-bandwidth use cases
- Thread for low-bandwidth use cases
- Bluetooth low energy (BLE) for provisioning of new devices

Here is a typical scenario of how everything comes together. When a new smart home device is turned on for the first time, it cannot connect to the local Wi-Fi because it doesn't know it. That's where the BLE comes in to securely have an administrator connect to the device (say, through a smart phone's bluetooth radio). It is secure because BLE only works over short range. Once that bridge is established, the user sets up the device to connect to the Wi-Fi network. From that point on the device can directly communicate through Wi-Fi. Low-energy and low-bandwidth devices, however, will need the so-called border router. These border routers will act as the translator between Wi-Fi and Thread. In this scenario, a number of low bandwidth devices will connect to the network through the border router that, in turn, connects to the Wi-Fi router. Note that the specification does not limit other means of communication with the devices. For example, if a manufacturer decides to connect with its devices over the cellular network, Matter doesn't limit that. What is required is that at least one of the above-mentioned radios exist. Figure 4.15 shows a logical view of the above scenario.

There are a few important security and privacy aspects to Matter certification. All the communication between devices or between a device and the cloud needs to be encrypted by AES 128-bit. That requirement strikes a balance between security on the one hand and compute power and latency on the other hand on lower-power and smaller devices. By imposing a certain security baseline, the certification encourages security by design as opposed to patch work afterward. Another important aspect of Matter is its focus on the

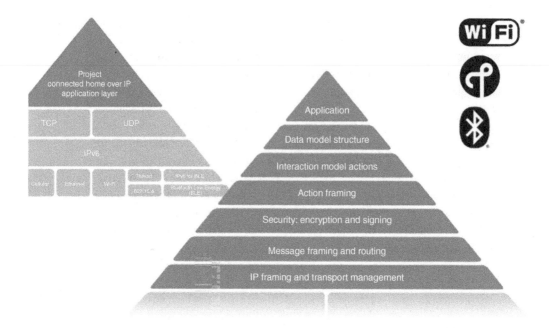

Common application layer + data model
Interoperability, simplified setup and control

IP-based
Convergence layer across all compatible networks

Secure
AES-128-CCM encryption with 128-bit AES-BC

Open-source development approach
Based on market-proven technologies

Common protocol across device and mobile
Extendible to cloud

Common data model
Core operational functions, multiple device types

Low overhead
MCU-class compute, <128 kB RAM, <1 MB flash

Figure 4.14 Matter formerly known as Project CHIP provides security bet practices besides interoperability (*Source:* Courtesy of Connectivity Standards Alliance).

enablement of over-the-air updates to both software and firmware. That way two security goals are achieved: on the manufacturer or servicing side security holes can be patched much faster and cheaply compared to the traditional customer campaigns or recalls. On the consumer side, it shifts the burden of keeping up with all such updates from the user to the manufacturer.

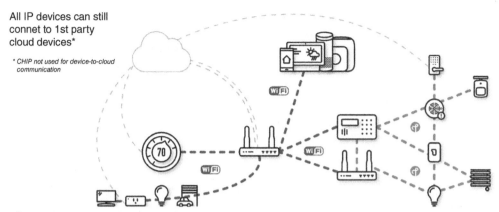

All IP devices can still
connet to 1st party
cloud devices*

*CHIP not used for device-to-cloud
communication

Figure 4.15 Matter's logical diagram (*Source:* Courtesy of Connectivity Standards Alliance).

Matter certification also provides a secure blockchain ledger to keep track of every single device's lifecycle. Imagine that over the years billions of IoT devices are manufactured, upgraded, disconnected, or destroyed. In the mix, some bad players may also inject counterfeit products under reliable brand names. The general consumer may not be able to track the software lifecycle of all of the smart devices in their home nor can be experts to tell a genuine device from a counterfeit. The blockchain ledger provides a secure and immutable way for the authorized parties to continuously create and update events related to every single device at scale. Events such as a firmware upgrade will be recorded in the blockchain, enabling the consumer to monitor and track the health of their devices.

Governmental Regulations and Other General Guidelines

In this part of the chapter, we review certain IoT-related regulations that are proposed or approved into laws in a few countries around the world. We will also recount a number of general guidelines that we propose any good IoT product should abide by in order to serve its users in a secure way.

So far we exclusively discussed private entities and consumers' efforts in making IoT secure: we discussed how hardware and software designers should build their systems in a way that makes them increasingly difficult for bad actors to hack. We also mentioned certain guidelines for consumers and consuming businesses when they choose devices and services of their choice. What is glaringly missing in this picture is the role of government bodies and regulators. It is no surprise that rules and regulations fall behind the cutting edge technology: social media long reigned the consumer data privacy domain until lawmakers gradually started to catch up and define guard rails; by some measures that domain is still work in progress. Other online advertisers started collecting and monetizing users' data for several years before solid measures such as the EU's General Data Protection Regulation (GDPR) or California's Consumer Privacy Act (CCPA) were passed into law. IoT is no exception. First of all, lawmakers need some time to see and grasp the scope of a technology in society before they jump in and regulate it. In certain, more innovative cultures any rule that is unnecessarily passed is considered as one more barrier in front of the innovators and is frowned upon. Also not all typical lawmakers live on the cutting edge of the technology. IoT and all of the nuances of its ecosystem take lots of effort to fathom. Due to all of these factors, the rules and regulations concerning IoT have been behind the technology by at least a decade. However, certain countries have started passing laws over the past few years to define and codify the lawful and recommended practices.

One such country to publish a Code of Practice for manufacturers and designers is Australia. The official title of the code that was published in 2020 by the Commonwealth of Australia reads *Securing the Internet of Things for Consumers*. In this short publication, 13 principles are laid out. They also published a separate guide for consumers and small businesses. In that guide, they briefly explain best practices before an IoT device is purchased, setting up IoT devices, maintaining IoT devices, and disposing of IoT devices. Critics argue that these publications, while useful, don't go far enough in describing the details. On top of that these are best practices and suggestions as opposed to rules and enforceable regulations. As an example, critics say, the UK also published similar recommended best practices without many results to show for it. Britain's Minister for Digital Infrastructure, Matt Warman, is quoted to say in July 2020 "Despite widespread adoption of the guidelines in the Code of Practice for Consumer Internet of Things Security, both in the UK and overseas, change has not been swift enough, with poor security still commonplace." Nevertheless, these are all good starts to a much needed set of regulations and consumer guidelines worldwide. We will briefly review the Australian version below.

Australia's published Code of Practice describes its intention in part as "The Code of Practice: Securing the Internet of Things for Consumers (Code of Practice) represents a first step in the Australian Government's approach to improve the security of IoT devices in Australia. This Code of Practice is a voluntary set of measures the Australian Government recommends for industry as the minimum standard for IoT devices." Manufacturers of IoT devices are encouraged to comply with as many as the 13 published principles voluntarily. By stating that in their product documentation, the hope is that the buyers will also play a reinforcing role through the power of their wallets.

1. **Passwords:** passwords are, not surprisingly, the first item on the list. The guide mentions that use of default, weak, duplicated, or easy-to-guess passwords are highly discouraged. Also, users should not be able to revert back to any default factory password. Two-factor authentication is also encouraged when sensible.

2. **Vulnerability disclosures:** manufacturers are encouraged to maintain a public place for users and researchers to share their findings about discovered vulnerabilities. Such vulnerabilities should be addressed in a timely manner.

3. **Software updates:** software updates, including third-party components, should be securely, easily, and regularly updated. On top of that such updates should be

smooth for the users in the sense that they don't change security or privacy settings of the user. Also, at the time of purchase there ought to be a guaranteed minimum support span for the devices during which users can be assured that the manufacturer will keep the software updated.

4. **Storing passwords:** passwords should be stored securely (encrypted) on a device without ever being hard-coded.

5. **Protection of personal data:** first of all personal data should be collected only when essential to the functionality of the device or service. And when such personal data is collected and stored, it should be secured in compliance with all the rules and regulations of handling personal data.

6. **Smallest attack surface:** any functionality or opening can potentially be exploited by attackers. Therefore, unused ports, function calls, or other features should be closed in the first place. Also, when handling privileges of identities, the least permissible amount of permission that is sufficient should be granted.

7. **Security of communication:** all communication to other devices, servers, or clients should be done in a secure manner. Besides, there should be a log of all such communications with the identity of the other party, as well as the time of the communication.

8. **Software integrity:** when a device boots, it has to check for the integrity of its software. At the moment that a check fails, the user or administrator should be notified; in the meantime, the device should not be able to communicate to other devices in the network to minimize potential spread of the risk.

9. **Resilience to outages:** one of the ways that hackers put IoT devices in a compromised state is manipulating its connection to the network or source of power. IoT devices should be designed and built in such a way that makes them resilient to those kinds of disruptions as much as possible. Peer or otherwise connected IoT devices should also be able to remain secure if certain parts of their peer network is offline or unreachable.

10. **Use of telemetry:** in case IoT devices provide data remotely about their state and usage, they should be constantly monitored for anomalies or signs of security breaches.

11. **Ease of deleting data:** users should have a transparent and easy way to delete their personal data and settings at will. If a user decides to discard the device or give it to someone else, they should be able to bring the device back to its factory state (without any trace of their personal data) with ease.

12. **Ease of use and maintenance:** usage and maintenance of a device should be easy. Oftentimes complexity in leveraging a device's features lead to poor or compromised state of a device, which can be exploited by bad actors.

13. **Validation of input data:** any input data through a device's user interface, network, or programmatic interfaces should be validated for authenticity.

These 13 pointers indeed make a comprehensive and valuable set of guidelines for any IoT product. The immediate next question is how can this daunting list be designed and implemented in practice. The tools and the required range of expertise are quite wide. Next, we will briefly review a design framework called *Zero Trust* that can help fulfill several aspects of a secure design. In a zero trust design model, the components always assume a request is a breach unless it is proven otherwise. Other ways of describing the framework is *never trust* or *always verify*. There are five key capabilities that are required to implement a zero trust security model.

- **Strong Identity:** devices and services always have to authenticate themselves in as strong a way as possible. For example, leveraging hardware signature authentication and recurrently renewable passwords are recommended. Organizations are also recommended to maintain an IoT device registry securely as the source of truth.

- **Least Privileged Access:** this is a fundamental principle that applies to software elements on a device, as well as network configuration. A service should be granted just enough privileges to get its job done. Any extra unnecessary power can result in an increased blast radius should it get compromised. In the same manner, networks should be segmented using some of the techniques we reviewed earlier so that the scope of any breach is limited as much as possible.

- **Device Health:** any device in the network should be monitored regularly for its health. Processes should be developed to apply hotfixes or label an unhealthy device accordingly. Leaving unhealthy devices in the network can have security ramifications.

- **Continual Updates:** all the devices should be updated regularly. This principle applies to both first-party components, as well as third-party. Out of date devices should prompt about the risks effectively in their user interface, so that the user is aware.

- **Security Monitoring and Response:** all the devices should be constantly monitored for abnormal behavior or any sign of being compromised. These tests range from simple rules to more sophisticated technologies that leverage analytics and machine learning. Also, penetration tests (pen tests) by white hackers should be part of the process in order to simulate real hacking scenarios.

Chapter 5

IoT System Design Process and Main Components

In this chapter, we take a deeper dive into several important concepts of an IoT system such as the design concepts, sensors, file systems, machine learning, and communication standards. By the end of this chapter, we have gained a broad insight into the major components that are needed for a typical IoT product. We discussed security in its own dedicated chapter due to its importance and volume of content; therefore, we will not discuss the topic in this chapter again. Needless to say, each project is unique in its own way and, therefore, the process and components may vary one way or another. Take the concepts of this chapter as the blueprint of an IoT product. We review technology, processes, and people aspects of the IoT product design. We also mention some aspects that are valuable for service providers in this space that may not necessarily be the same as the manufacturers of the IoT device.

DESIGN AND DEPLOYMENT PROCESS

When it comes to the IoT systems, like most other engineered systems, the requirements need to be specified first. The scope of this chapter is mainly the process of building the product; therefore, we assume that the market analysis and the generalities of product specifications are handled beforehand per the other chapters of this book. After the specification is complete,

IoT Product Design and Development: Best Practices for Industrial, Consumer, and Business Applications, First Edition. Ahmad Fattahi.

system architects build a high-level architecture of the system that highlights the bird's eye view of the system and its components. Engineers and designers tackle various components of the product and design material, and mechanical and electronic components of the system to be prototyped. Quality assurance professionals test each component individually and while integrated with other parts until all tests pass. The production begins then.

To break down the major phases in a top-down way, we distinguish the following specific steps. The level of details and fine grained engineering nuances increase as we proceed down this list.

Product requirements: Everything has to start with requirements that explicitly document the must-haves in clear terms and in writing. It can come from customers, sales, marketing (competitive analysis), management, or engineering teams who learn from failures or are just innovative. A very important aspect of this phase is that the ownership of requirements needs to be very clearly defined. Product managers often take on this role who then get advice from many other sources:

- Customer research and success teams: these teams directly research, organize, and translate the wants and needs of the market as a whole, as well as granularly, and various personas in the customer base in a way that leads to specific products. For example, a market and customer research team can conclude and recommend that health monitoring features are a major category of wants by smart watch customers. Or in a manufacturing environment, smart helmets that can provide audio information to plant workers is another example.

- Market and business development research: these teams are the equivalent of the customer research teams but at a macro level. Such professionals look at analyst reports and market trends, as well as competitive and adjacent markets and propose macro trends in specific markets. An example is how enabling smart solar energy through IoT will have a bright future due to market trends and public sector subsidies. These requirements should include hard opportunity analysis with objective numbers such as total addressable market, best and worst case scenarios in terms of market share, as well as objective competitive risk analysis.

- Support and field teams: these individuals are at the forefront of interacting with customers to solve their issues or integrate products for them. As such they often have a very unique and valuable angle to what works and what is needed to improve in the product. Mature and smart organizations have well-thought processes to channel and absorb the feedback from these teams into their product design teams. These processes can be a combination of human testimonials and artificial intelligence-powered tools that extract topics and insights from large bodies of field and support reports.

- Engineering teams: the process of building the product is a highly innovative and challenging process. Engineers often start thinking outside the box and come up with innovative new ways of building the product that improve the user experience or the economy of the product itself. A point of caution, though, is that this aspect of influence in the product design should be delicately balanced with the ownership of product managers. If the pendulum swings too far in the direction of engineers, the results may be genius works of innovation and engineering that few actual users want to use.

- Security professionals: given the ever-increasing importance of security as a major aspect of any IoT product, the security experts have to have a say in the requirements process. For example, if an IoT product handles sensitive personal health data, logging into the product website may require two-factor authentication by default. Compared to the other groups with recommendations, security professionals may have a niche area of interest. However, it is critical that they are at the table and deeply involved from the beginning in both this phase and the next phase of architectural design. Making security an essential main ingredient from the beginning is a must.

Architecture Specification: the next level of detail is designing the high-level architecture of the product. A good architectural design includes the high-level description of the flow diagram of major components and guidelines about their specifications. It, however, stays away from deep detailed instructions on how to engineer the system. For example, an architectural specification document may recommend that the product includes a communication module that supports Bluetooth Low Energy for device provisioning and setup. It also specifies how this module connects and communicates with the rest of the system. A mechanical architectural specification, similarly, defines and architects the form factor of the product. A common issue is when the architecture ends and engineering starts; drawing the line can be subjective. As a rule of thumb, as long as you *believe* there is a reasonable engineering solution to the architecture design, stop there. Don't go beyond that point in the architecture design or you risk crossing over into the engineering territory. Visual representation of the architecture specifications is highly recommended because it helps engineers build the right design.

Developing Components: building the individual components separately per the design specifications is the next step. In this phase, details need to be decided and heavy engineering work is done. For example, imagine that the architecture and the requirement specifications call for a machine learning module that can detect certain images with AI models. This component can be carved out as a standalone component to be engineered and prototyped by a team of hardware and machine learning engineers. Note that the architectural design specifies on a high level how this module will interact with other modules such as the camera, main device memory, and power. It also defines how much power can be consumed by this component and how large can the physical dimensions be in a way that does not violate the overall specifications of the product. This is when the most detailed and often challenging parts of the work happens. Lots of mechanical, software, and hardware engineering work is broken down into several teams that aim to build and deliver the design per the architecture design document. Oftentimes, as the engineering work progresses, various teams lobby for more legroom in their design or more engineering resources in their teams to make their deadlines. A real-world example was when the facial authentication on the iPhone had design specifications that would allow the user to register more than one face; at the same time, the facial recognition for the user should have happened under a specified fraction of a second by the device. These two design decisions put the machine learning and hardware teams under a lot of strain resulting in lobbying for a reduction in the number of registered faces; they were successful and won the argument to curb the maximum number of registered faces in the interest of faster user experience.

System Integration and Testing: modules and components that are built individually are put together next. Never underestimate the importance of this phase because many things can go wrong when things connect with each other. Mechanical and electromagnetic interference by a component can interact with the operation of other components that in isolation work perfectly fine. This is where thorough testing is utterly important.

The above steps are almost never completed in a linear manner in reality. There are several adjustments in the upstream phases when the downstream makes discoveries or finds issues. Also, the physical and economic limitations can pose new realities over time that require upstream changes. A well-designed process with clear roles and responsibilities and well-staffed teams are essential to the success of an IoT product.

Each team size is a function of project complexity and its ROI. The immediate question is how a project's complexity should be gauged. One way to measure the complexity is to break it down according to scope, schedule, and skills.

Scope measures how sophisticated and pervasive the hardware and software components are. Another way to look at it is the number of features and their complexity level. You need to ask how many features from the prior products can be transported to save time and resources. On the other hand, you need to look at how much new work needs to be done. Engineering intuition and experience will be key to making the right judgment.

Schedule depends on the expectations set for external personas, such as customers and partners. Product teams define and publish an internal calendar that needs to be observed. Don't be alarmed if you see a heated debate between the engineering and sales teams. While Sales wants to sell more faster, the engineers may prioritize meticulous attention, quality, and their own work–life balance. It is very common to see the tug of war between the technical and business sides of the organization when it comes to the calendar of releases. It needs to be taken seriously and constructively in order to find the sweet spot between the factors mentioned above. Ownership of the calendar matters a lot so that it is clear to everyone where the buck stops.

Skills and Processes should be the right fit to the project at hand. If, despite having an adequate number of staff members, the project is falling behind it may be suffering from lack of process or the right skill sets. Adding more people (some people call it throwing people at a problem) doesn't necessarily translate to faster execution. In many cases if more people are brought in without a proper process, actually the team can end up slowing down. The right allocation of resources is also a function of time. Figure 5.1 shows four skills on a chart versus time. The area under the curve is the operating cost of delivering a product. A typical breakdown of the product team can be seen below. Each project will have its own nuances that may need a different complexion. However, the general concepts

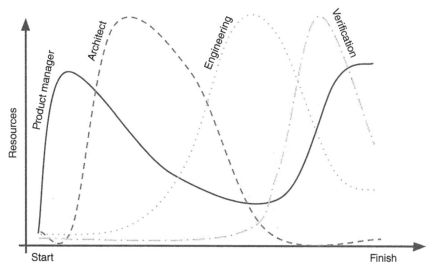

Figure 5.1 Project management heap chart.

can be safely generalized. A subset of the engineering team is saved for verification engineers. Focusing only on builder engineers can result in systems that are either hard to use or are filled with bugs. Verification engineers can kick the tires on features and also try scenarios that design engineers may not see easily. In real IoT teams, verification teams make up a significant part of the total workforce, which is a result of the complexity of these products and potential use cases.

Roles, Responsibilities, and Team Structure

Now let's talk about specific roles in the team in more detail (Figure 5.2). The software architect is the highest authority on how the software needs to be designed. She knows different technologies their pros and cons very well and has a clear vision on how different

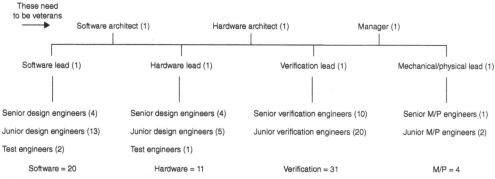

Figure 5.2 Project team break down.

components do or do not integrate well with each other. The hardware architect is the highest authority on how the hardware needs to be designed. He has experience in chip design, as well as physical and mechanical considerations for the form factor. He has a vision on serving software needs with the hardware platform. You need a manager that acts as the glue, owner, and the bottom line arbiter in technical and scope disputes. The manager is also the guard to the product vision and the judge when to ask for extensions or feature adjustments. The design team comprises the following main verticals.

The software team is in charge of building the software components on the provided hardware. The team is typically led by a technical lead. This team can meet in a daily stand up meeting to discuss their progress. They are very well integrated and work as one unit. Software test engineers make sure that the different components in the larger software system are built correctly and no major bugs are shipped. The size of the software team is significantly larger than the hardware team. While this can vary based on the project, the general trend in the industry agrees with a lop-sided balance tilting in the direction of the software team. Programmable chips make it even easier to start from a generic hardware design and then apply the desired software at the chip level. The hardware group also needs its own technical lead. Other aspects of this team mirror those of the software team.

The verification team is usually the largest of the three. This is a result of the complexity of the resulting system when individual software and hardware components are integrated with each other. Unit tests need to be designed and performed in a systematic way that covers several different scenarios. The cost of shipping a product that is not properly secured or suffers from an abundance of bugs can push a small business into bankruptcy.

Most IoT projects encompass a physical form factor including casing, heat transfers, cable and connector design, and similar mechanical components to make sure the signals do not interfere with each other and the usability is optimized. The relatively small mechanical team is responsible for this aspect of the product.

It is crucial to define, publish, and reinforce the roles and responsibilities of everyone in an IoT team. A perfect process and the right people will still be inefficient and chaotic if the individual team members are not clear on what they own or what role they have in the process. There are various models to build a good process. One such model is called RACI for short. This model defines four possible roles for anyone associated with a project or task:

- **Responsible:** this is the person or group who is responsible for executing the tasks. For example, an engineer who is assigned to write a piece of software that detects anomalies in sensor data in two weeks is considered responsible for this task.

- **Accountable:** this is where the accountability lies. Usually team leaders and managers assume this role in projects. They will have the power to assign tasks to individual team members (or matrix team members) in order to be held accountable. If deadlines are missed, higher levels of management will hold this role accountable.

- **Consulted:** these are subject-matter experts or adjacent teams who may be able to inform decisions. For example, a mechanical engineer may be consulted on how much heat a processor can dissipate before it overheats. That consultation will then lead to electrical and power decisions by the hardware team.

- **Informed:** downstream and adjacent teams who may be impacted by decisions in the project are kept informed. The rule of thumb is if a certain decision impacts another team's own decisions, they should be at least informed. For example, when a significant feature is added to the product, the marketing team should be informed about it.

By now we have established a framework for various roles and responsibilities in the IoT project. Let's map the core functions in the example chart above to the RACI structure. The manager wins business buy-in, defines the roles and responsibilities, sets the vision, and brings in the right resources. She makes sure that the design reviews are done properly and people performance reviews are performed as expected. In short, the manager is the guardian of the whole project. The manager is accountable for setting the goals, repeating it at the right frequency, and bringing in the right resources.

Hardware and software architects are accountable and responsible for the high-level architecture of the product. The hardware architect holds the software architect as consulted on hardware issues and vice versa. Due to this accountability, they are the highest level of technical decision in the team. Senior design engineers and leads are accountable for designing their respective components per the architecture. They produce detailed documents on how each part of the architecture is to be built in technical terms. Oftentimes, the architects consult the lead design engineers or at least inform them during the architectural design process. Individual engineers are responsible for implementing the solutions while informing the adjacent and impacted teams of their implementation decisions. Certain disputes can get escalated to technical leads or architects when requirements collide. Test and verification engineers are responsible for the quality assurance and testing of the product after they produce a document detailing the test plan. The verification leads are responsible for the general test plan and also accountable for the overall verification process of the product. They remain informed, when it makes sense, by the product engineers about the details of product implementation. And finally the mechanical and physical team is accountable and responsible for mechanical design of the product, as well as its thermal, shock, and vibration analysis.

Process – A Deeper Look

In this segment, we take a more detailed look at how the design and deployment process works in IoT product teams. Processes are required to minimize the friction between people and between people and technology. Its eventual goal is to minimize the area under the heap curve of functions while achieving the product goals. In other words, it optimizes the resources while the goal of the project is achieved. All good processes in product and project environments start from a solid set of requirements before the design, architecture, and the development start.

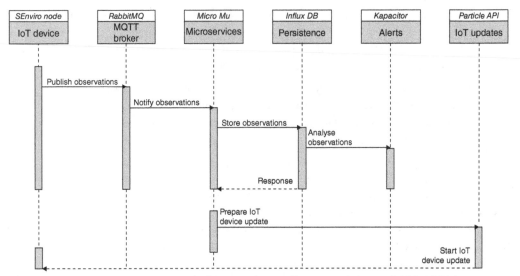

Figure 5.3 UML diagram example (*Source:* https://www.mdpi.com/1424-8220/20/8/2418/htm).

The architecture definition process is in many cases a top-down process. The architect authors the requirements document with clean deliverables. The main features, cost, power, size, weight, and performance are specified here. The product managers produce a document that is written in user language. They talk about the land of customers and users. The architecture document absorbs the product definition from product owners and defines what the product does in technical terms, how it works, how data and control flow, and everything else that a technical lead needs before they can start building the components. One of the ways architects communicate their architectural designs is through UML diagrams. These diagrams clarify how various pieces of the product are interconnected. They also show how data and control flow as time elapses on the vertical axis. Figure 5.3 shows an example of a UML diagram that engineering leads can study, debate, provide feedback, and eventually start implementing with their own teams.

The example UML chart shows why this method is top-down. It builds and imposes an overall schedule for how different parts of the system jive together. It also clearly defines how performances are measured for each part of the system. For example, if a software service call is made to another process, the architectural design defines the maximum allowable latency. Exit criteria for the architecture phase into the engineering phase are very important. These criteria allow the architects and designers to consider the design solid and stable enough for the more hands-on engineering team to start building a plan around it. Absent of such exit criteria, architects can keep detailing aspects of the design to lower levels of specifications for a long time. Once the exit criteria are met, significant changes to the high-level design may not happen easily; the design is considered frozen so that the builders can start building.

The architectural design may also happen in a bottom-up fashion. In this methodology, product owners and the leadership engage with various vendors and consultants to get specifications for various components of the product. After the selection is made, detailed specification documents for software, firmware, hardware, data and control flows, permissions, test plans for each module, and other aspects of the product are made. This process is much heavier up front on the engineering side. These documents are typically made by the engineers and based on the modular designs received from various vendors and consultants. The upside of this approach is that the builders are the actual teams who do the upfront design. Therefore, moving from design to development is often much faster than the top-down approach. However, the challenge of this approach is to make sure that the different components eventually integrate well with each other. It is absolutely critical that specifications for the test bench are crafted at this point and before actual development has started. Everyone should know in advance how the design will be verified. This verification plan needs to clearly describe how the test bench should be built with all the inputs, outputs, scenarios, and pass/fail criteria.

Once the individual designs are in place and test plans are approved and communicated, a project manager builds a real schedule from bottom up and delivers a realistic timeline. The project manager is very helpful here to work with the designers to collate a full schedule that is much closer to reality than the top-down approach. Project management tools such as Microsoft Project or Jira are widely used in this phase. Similar to the top down approach having clear exit criteria to kick start the development phase is a must. Usually higher-level leadership and stakeholders need to approve the end of this phase by green-lighting the design. At that point, the design is considered frozen. Applying changes in the future will be harder and will have to go through a separate process.

Once the design is approved and frozen the team enters the *development phase*. This is the phase where software, hardware, and physical development happens. Not surprisingly a significant amount of money and resource expenditure is required for this phase that can be one to two orders of magnitude larger than the design phase. The development teams should be religious about following the approved design document; if ambiguities arise (which is not uncommon) they should raise it to the leads and architects. The mantra throughout this phase is that knowing *earlier is much better than later*. It means that any issue in the design or development that remains unknown and the longer it remains hidden, the more expensive the fix becomes. As we see later, it is especially true if a design moves to mass production or, even worse, ships to customers. As a corollary, the full architecture and design plans need to be complete and solid before development begins. In many ways, making a U-turn or taking a detour keeps getting costlier as the project progresses.

It is also important for the leads and architects to anticipate unforeseen issues both at the design and development levels that can hold up the whole agenda. While they need to expedite the resolution they will need to remain calm and spread the sense of control to their corresponding teams. Having evergreen projects or optional improvements available can be handy to leverage any lingering capacity in the team. The exit criteria for each development phase is defined as matching the specifications in the documented specifications in the design document; the criteria have to be very thoroughly followed.

Once individual development is done and the exit criteria are met, *manufacturing* starts. Note that this phase is still one step before the mass production. The developed set of modules go into manufacturing for the testers and verifiers to actually vet them in flesh as opposed to computer simulations and models. Many IoT products include custom chip design and manufactured components in them. Application Specific Integrated Circuits (ASIC) is a common technology to go about such custom chip designs. The common Register Transfer Level (RTL) description of the hardware is developed by the hardware engineers, which will eventually get translated to low-level code that drives the IC manufacturing machines. All such code generation practices speed up significantly if the prior phase design documentation is done clearly and thoroughly. Also included in this phase is the physical design of the ASIC as to where each pin and cell should be placed while the automatic test programs verify the quality of the produced chips.

Before meeting the exit criteria from this phase, a well-designed test and verification process needs to be performed. Once again an emphatic reminder of the mantra: *knowing earlier is much better than later.* Bugs and unmet specifications are much costlier if they slip through from development to mass production; in fact the difference in cost can be an order of magnitude. Smaller startups can easily cease to exist in face of flawed products that go into mass production. Under pressure, employees, engineers, and testers may overlook flaws or questionable results intentionally or out of fatigue. They need to be actively encouraged by the leadership to watch for sneaky bugs and report them. Once that is done the good deed needs to be publicly applauded. A well-designed verification process entail the following three concepts:

- Coverage refers to the scenarios and aspects that will be tested. Some IoT use cases are simple and some can be extremely sophisticated. Regardless, all line items in the design description need to be well tested. This is especially important in more complicated cases where several combinations of operating modes and functionalities may exist and need to be verified. Corner scenarios and real-world situations all have to be closely considered and listed. Automation is the verification engineers' best friend in this context to take care of a large number of permutations.

- Vectors translate the covered scenarios into the inputs that go into the chip to test each of the scenarios; in other words, they are the test scenarios stated in a lower-level description. Note that depending on the complexity of the design, mimicking the high-level condition that is desired in the coverage may take some well-thought thinking and engineering work. It may entail some static inputs or a sequence of varying inputs that simulate various conditions. The bottom-line goal is to test the manufactured chips and components in as close of a situation as possible to the covered cases.

- Checkers describe the readings of signals that eventually reveal if the test result is a fail or a pass. In other words, these are the outputs of interest for each test scenario that are the telltales if the test is a pass or a fail.

There are different approaches or methodologies to tests. In a direct test the tester explicitly specifies the test case and the checkers to verify the functionality of interest. In contrast to that you may run a constrained random verification test. In that approach, a random generator is used to simulate many combinations of inputs and pass them all through the test bench. The tester then looks inside the set of tested scenarios and picks the ones of interest to check the outcomes (checkers).

The results should finally be summarized in human readable reports for the verifiers. Rolled-up versions of the test results should also be prepared for managers and higher up stakeholders with meaningful KPIs to have full visibility into the coverage and test results. There are rigorous approaches to verification in many hardware design platforms. A great example of that is the Universal Verification Methodology or UVM where several aspects of the test can be automated.

The emphasis and investment in the verification phase is partly a result of how costly undiscovered bugs at this stage can be. When a design is green-lighted into mass production changing the mask design for the wafers can easily cost millions of dollars. You absolutely want to invest enough in the verification phase to avoid such catastrophic mishaps. If an issue is passed to shipped products not only the monetary cost goes up dramatically, the reputation and brand of the company may suffer as well. That's why knowing earlier is much better than later.

The exit criteria for this phase include matching the requirements in the design documents at a high-pass rate for the manufactured chips. The definition of high-pass rate may depend on the use case and the company itself. If the use case is for medical purposes or otherwise regulated industries, usually these thresholds are north of 99% and are encoded in the regulatory documents. For less-sensitive use cases or where the business approach is to compete on price, product managers may decide to lower the pass threshold to 98 or 95% in order to reduce the upfront production cost. Moving forward from this stage requires very high-level approvals. The reason is that having the masks produced for mass production is quite an investment for any company. The (randomized) verification process may not stop even after the mass production begins. Sometimes as the chips age for several more days or the environmental temperature changes, the tests can hit unique scenarios that have not been tested before. Depending on the issue, sometimes mitigations can be accomplished even after the design is sent for mass production; the likelihood of such mitigations decreases as the process advances further.

Finally, the approved developed components go into the deployment phase. The test and verification practice continues here because of potential interaction issues among components or problems during the deployment process. This is to make sure that the chips and the whole product work as expected at a high rate. In case bugs are discovered, a list should be created systematically and passed over to the engineers for future improvement. There are machines in the market that can run test scenarios and simulations much faster than the actual design tools; using such tools, if necessary, can speed up the validation process. Some companies build prototypes or qualification units where a full product is shown to a small group of trusted or randomly selected users for feedback. After a certain point in time, the discovered bugs or enhancements will be moved to a backlog for future releases.

Security is not to be forgotten! Parallel to all the features and usability validations, security tests make an important part of the test scenarios. There are professional security testers in the market whose job is to hit the prototype with a list of known scripted attack scenarios. Their value doesn't stop at their security expertise; they are also valuable because they are external to your company and potentially bring fresh eyes to the conversation. As we discussed in the chapter on security, security takes an orthogonal way of looking at things that is often missing in the builders' circles. Having fresh eyes trying to break the product is a good idea.

This phase is a full in-form-factor set of tests that validates the product functionality exactly as if an end-user would. Prior earlier test stages are usually done on individual modules out-of-form-factor. It can potentially reveal brand new bugs. It can be because of overlooked bugs or, more importantly, due to the interaction of different physical and software components of the product for the first time. Apple's so-called Antennagate in 2010 caused many enthusiastic fans of iPhone 4 to be disappointed when they saw their phone antenna coverage drop unexpectedly. Although there were some serious design flaws, a full validation process with in-form-factor scenarios would have caught this issue before the whole company was embarrassed.

One of the common ways to validate products is called burn-in. In this approach, a number of fully developed products are run for a large number of hours or days in controlled environments. That way statistically many more flaws will surface. It can be combined with exposure to various environmental conditions such as temperature, humidity, lighting, motion, vibration, electromagnetic interference in the ambience, and other unforeseen conditions. If these conditions are pushed to the extremes, it is often called stress testing. Burn-in testing is also a good measure of how the product generally ages. Bugs that are revealed through this kind of test are often classified into infant, steady state, and wear out issues depending on how early they appear. The corresponding time span of this test should be designed relative to the life expectancy of the product.

Although we mentioned a few physical factors to consider for testing, it is worth calling out temperature because of its pervasive impact on most products. When the product's ambient temperature changes, both its physical and electronic characteristics change. Tiny metal connectors in the circuit boards expand and contract, which can lead to breakage and loss of connectivity. Alternatively, insulators can crack leading to unwanted electrical connectivity. The characteristics of the wafer also change by temperature leading to different electronic behavior. Needless to say, depending on the targeted market and intended use case, the accepted temperature ranges vary significantly. Some of the more common defined ranges for products in various sectors are listed below:

- Full military: −55 to +125 °C
- Automotive: −40 to +125 °C
- AEC-Q100 Level 2 (extended automotive use cases): −40 to +105 °C
- Industrial: −40 to +85 °C
- Extended Industrial: −40 to +125 °C
- Commercial: 0 to +70 °C

We don't get into too much more of the details of thermal testing. It suffices to say that thermal testing is very important and that even the process of designing the test is not trivial. The rate of temperature change and where to position the source of the heat can have important ramifications for the test. Good test schemes also subject the product to a matrix of various temperature and voltage combinations. Test and validation engineers are trained to perform effective thermal tests. Once these tests are satisfactorily passed mass production and go-to-market starts.

SENSORS

Sensors are ubiquitous in IoT systems. It is hard to find an IoT system that does not sense things around it. Any smart system needs to react to the situation around it, which means it has to be able to sense physical quantities such as light, temperature, pressure, voltage, acceleration, or a myriad of other quantities. In this section, we first review some of the more common sensor types in IoT systems. Then we turn our attention to some of the general concepts that define and qualify a measurement system such as resolution and accuracy.

While some sensors are simple in that they measure only one single quantity (such as temperature), some other ones are more sophisticated. For example, a specific motion sensor may have a light sensor or even a camera inside it that feeds its reading to a small analytical engine. The engine sifts through data for a specific type of pattern that is the signature of the motion of interest. While modern applications are digital, some of the older industrial appliances, such as pressure gauges, may be analog. In the latter scenario, the IoT sensor leverages some secondary or proxy measures to turn the analog reading into a digital signal.

Common Sensors

Given the importance of sensing to an IoT system, there are various types of sensors available to measure several types of quantities and events. Many factors play a role in choosing the type of sensor. For each type of quantity, for example temperature, there are various technologies that offer their own pros and cons. Other than price, availability, ease of use, and other business factors each sensor has some technical specifications that we will review later. For example, a sensor may not be fine enough or inversely too granular of a measuring stick for the use case of interest. Sensitivity to changes in the ambience, motion sensitivity, or how fast the sensor ages (drift) are among other factors to consider. Going into every such factor for every type of sensor is beyond the scope of this book. But we will spend some time covering the more common concepts.

Temperature Sensors are some of the most widely used types of sensors. These sensors turn the temperature into electrical signals that can be turned into digits and communicated to the rest of the IoT network. Applications in consumer use cases abound and include home thermostats, water heaters, and smart watches with health features. Industrial use cases cover several different areas of manufacturing and process engineering, automotive, and agriculture industries, among others. Below are a number of common types of temperature sensors.

- *Thermocouples* are essentially two pieces of wire and a voltmeter. When the temperature changes the voltage difference between the two ends of the thermocouple changes accordingly, which enables the device to interpret the sensed voltage into temperature.

- *Thermistors* are thermal resistors. They are resistors whose resistance varies with temperature more rapidly than typical resistors. Therefore, a circuit that can measure resistance can map the sensed resistance to the ambient temperature. Some thermistors have a positive thermal coefficient (resistance increases with increasing temperature), while some other ones have negative coefficients.

- *Semiconductor* thermometers leverage the varying resistance of semiconductors with temperature in an integrated circuit and directly read the temperature in digital format.

- *Infrared* thermometers can sense the temperature of a solid or fluid from a distance without touching it. They infer temperature from the amount of infrared energy that is emitted from an object, which is a function of its temperature. Figure 5.4 shows a commercial infrared temperature sensor.

Humidity Sensors detect the amount of vaporized water in the air. They matter in industrial use cases because many processes such as food and beverage, materials, chemicals, and others rely heavily on an accurate trajectory of humidity during

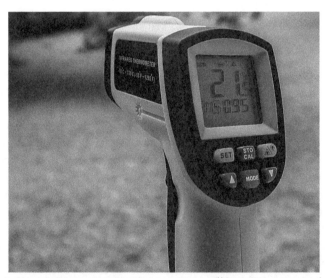

Figure 5.4 Infrared thermometer (*Source:* https://en.wikipedia.org/wiki/Infrared_thermometer#).

production. They are also important in consumer applications both for health and comfort reasons. Usually humidity sensors are one of the three types.

- *Capacitive* humidity sensors measure relative humidity through a proxy capacity variable. It places a thin strip of metal oxide between two electrodes. The metal oxide's electrical capacity changes with the air's relative humidity. This type is mainly used in industrial and meteorological applications.

- *Resistive* humidity sensors utilize ions in salts to measure the electrical resistance of atoms. As humidity changes, so does the resistance of the electrodes on either side of the salt medium.

- *Thermal* sensors turn humidity into a difference in measured temperature and then measure the temperature difference. They use two thermal sensors; one sensor is placed in dry nitrogen while the other measures ambient air. The difference between the two measures the humidity by proxy.

Pressure Sensors turn the pressure in a fluid or gas to an electrical signal. Pressure is force per unit of area. These sensors are commonly used in automotive industries for various aspects of a car operation. They are also widely used in process engineering where an ideal reaction calls for a closely monitored pressure in reacting materials. Health sciences also care much about pressure in ventilators, as well as the familiar blood pressure of patients.

Proximity Sensors are used to detect when an object is in the vicinity of a location. They are widely used in supermarkets, parking lots, and automated assembly lines. There are four common types of proximity sensors.

- Inductive proximity sensors use the change in the inductance of a coil when an object's position is within the range of the coil's electromagnetic field. This change in induction is converted into a voltage that in turn activates a switch for detection.

- Capacitive proximity sensors are similar to inductive sensors but rely on the change in the capacitance instead of inductance. Since almost all materials are dielectric these types of sensors can detect a wide range of objects.

- Ultrasonic proximity sensors emit sound waves at high frequency (20 KHz or higher) and detect their rebound. Given the relatively lower speed of sound compared to light these sensors can often sense distance to the object as well. Another good use for the ultrasonic type is sensing transparent objects. Note that many animals can hear 20 KHz and may find it annoying. Care is advised when pets are roaming.

- Photoelectric or opto-electronic sensors send a light signal out and sense the returned pattern. Through the analysis of the returned signal, they can sense if an object is proximate.

Level Sensors measure the level of a fluid or granular substance (such as powder or grain) in a container. They are abundant in food and beverage industries, as well as oil and gas, automotive, process, water and water treatment, and flood warning systems. One type of these sensors return a binary signal whether or not the level of the substance is above a certain threshold. A second type of these sensors returns a reading of the actual level of the substance as an analog or digital signal.

Accelerometers sense the rate of velocity over time. Their use cases range from smart phones and watches that track the users' motion to athletic equipment that monitor athletes' activities for performance optimization. These sensors are also very useful to detect motion in security and safety systems. The sensing mechanism is often one of the two types. Capacitive accelerometers have a capacitor in them, one of whose conductive plates can move with acceleration. Therefore, its capacity varies depending on the acceleration which, in turn, can be turned into a readable electrical signal. These sensors are usually more suitable for lower frequency accelerations. The second type is the piezoelectric sensors that use microscopic crystal structures to convert acceleration into voltage.

Gyroscopes measure the angular velocity of an object. They can also be used to show the *down* direction in gravitational fields. These sensors can show gyration around one, two, or three axes. Often combined with accelerometers, they can provide a full reading of the positional situation of an object. Their use cases range from the transportation industry in airplanes and cars to personal smart devices such as phones and athletic trackers. Their technology is often one of rotary or classic, MEMS, vibrating structure or optical.

Gas and Smoke Sensors sense the presence of specific types of gas such as carbon monoxide, hydrogen, ozone, or oxygen to name a few. These sensors are widely used in residential monitoring systems, as well as factory plants and mines where monitoring the presence of certain poisonous gases is critical. Smoke detectors are sometimes used or combined with gas sensors to detect particles of certain size and properties in the environment. They are used to detect fire or other potentially dangerous conditions.

Optical and Infrared Sensors use visible or infrared spectrum of light to sense characteristics of their environment. In some use cases, the transmitter and receiver of the light are both in one device; an example is the blind spot detector of cars where the sensed reflection of the emitted infrared signal will determine if an object exists in the blind spot or not. In other use cases, the source and receiver can be parts of the same system but in different places; an example of that is most short range remote control systems such as consumer TVs and air conditioners. Touchless thermometers only have infrared receivers that infer the temperature of an object based on the level of infrared light emitted from it. Visible light sensors are also widely used in smart buildings and security systems. Many light detectors do their job by using a photoresistor. The resistance of a photoresistor changes by the amount of light it receives. A simple electric circuit can translate the varying resistance to a corresponding voltage that can be read and used.

Image Sensors are what enable the digital cameras in our phones. They use integrated circuitry (often CCD or CMOS) to convert optical signals into electrical signals on a grid. The resulting response forms the raw "image" which is then processed and interpreted as a visual image or print for the use of our eyes. Their use is common in security systems, consumer digital cameras, and advanced automotive features and self-driving cars.

Touch Sensors are used very widely in touch screens on smartphones and other smart screens. One of the more common technologies behind them is measuring capacitive change. In other words, they detect the change in capacity of a capacitor under the surface of the display when a part of the human body such as the fingertip is close to it. That capacity is in turn converted to a voltage response that can be used to interpret the touch. These types of sensors are very widely used because they do not require physical pressure on the surface of the screen.

General Sensor Concepts

We saw a number of common types of quantities for which efficient and economical sensors exist in the market today. To choose the right kind of sensor for the specific application and to integrate them properly in the rest of the IoT system design, we need to know several concepts that apply to sensors in general. On the one hand, we need to make sure that the chosen sensor provides the necessary amount of precision, resolution, and alike for the application. For example, most consumers expect their home thermostats to be able to control 1 °F or 1 °C at a time. A piece of lab equipment may require that resolution to be 100 times higher to 1/100 of a degree. On the other hand, choosing unnecessarily high-quality sensors can lead to increased cost and development complications without any return. For example, embedding a temperature sensor that distinguishes between 77.02 and 77.04 °F is meaningless for a regular residential thermostat, but it can increase the production cost significantly.

In our examples above, we were concerned about the *resolution* of a sensor. The resolution measures how fine-grained a sensor is in reading the quantity that it is reading. For example, if a pressure sensor can read the pressure to the 1/10 of N/m^2, its resolution is 1/10 of N/m^2 (or Pa).

The next immediate question is if we repeat measuring the same quantity with a sensor, do we see the same results. This question refers to the *precision* of a sensor. If a temperature sensor with 1/10 °C resolution is connected to an analog display that is only 1 °C precise, the whole combination's precision is 1 °C, and not better. Another way of looking at precision is how repeatable the process is: if the analog display can be trusted only to 1 °C precision, by repeating the same measurement over and over, we get the same result up to that level of precision. The finer resolutions in the process, in this case the digital sensor, become irrelevant.

Accuracy measures the correctness of a measurement system. It often gets confused with precision. To see the difference note that each measurement is subject to random noise in the physical environment. Therefore, two measurements of the same quantity almost always leads to different results. In other words, the resulting reading forms a distribution around the actual value in the real world. The accuracy is concerned with the center of gravity or mean of the distribution; if it is close to the real-world value, the sensor is said to be *accurate*. If the distribution is tight around its mean value, the sensor is said to be precise. A measurement system can be quite precise in that it repeatedly outputs more or less the same result. However, the result can be accurately or inaccurately distributed

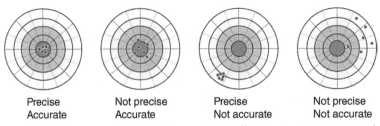

| Precise | Not precise | Precise | Not precise |
| Accurate | Accurate | Not accurate | Not accurate |

Figure 5.5 Precision vs. accuracy (*Source:* https://upload.wikimedia.org/wikipedia/commons/3/37/Accuracy-vs-precision-nl.svg).

relative to the reality of the world. Alternatively, a measurement system's outputs can be distributed accurately around the real-world quantity while the distribution is tight or wide. In other words, precision and accuracy can vary independent of each other. The tighter the distribution of the output values, the more precise the sensor. The closer the (average) results are to the true value of the world, the more accurate the system. Figure 5.5 shows the four permutations of a sensor in regards to its accuracy and precision.

Tolerance is a measure of how far off the results can be from the expected results. In some ways, tolerance combines accuracy and precision together. Any physical system and in particular sensors are produced to satisfy a certain level of accuracy and precision. However, nothing is perfect. Both manufacturing randomness as well as ambient factors at the operation time, such as temperature, lead to deviation from the nominal specification. Tolerance measures the distribution of a sensor's results around the nominal specifications. Therefore, an IoT designer uses tolerance to decide if a certain sensor can meet the desired specifications of a function when it comes to errors in measurement. Tolerance of a sensor states how much measurement error is expected to be tolerated. The tolerance band of a sensor itself is a function of the tolerance band of the components inside it such as resistors, or reference voltages. The tolerance is usually expressed as a percentage number. It indicates the difference between the actual value of the component or measurement with its nominal specification. For example, a reference 2.4 V reference voltage with 1% tolerance can be anywhere in the range of 2.376 and 2.424 V. Note that narrower tolerance bands can lead to significant increase in a component's price. Business justification needs to be applied to the design process when higher precision and accuracy components are chosen.

Any sensor needs to be calibrated and validated before being shipped out. Alternatively, some industrial users want to validate purchased sensors before putting them in production. Although each physical quantity is different we will continue considering a temperature sensor to introduce the calibration process. A very common way of building temperature sensors is using temperature-varying resistors or thermistors. The temperature-sensitive resistance can fairly easily be measured in runtime using first-principle circuit laws and a voltmeter. The remaining gap is to go from resistance to temperature. The relationship between the temperature and resistance is highly nonlinear. One of the common physical models that describe the relationship is called the Steinhart-Hart equation; it is a nonlinear equation with three parameters. Once the three parameters are measured for a thermistor, its behavior is considered known. The question is how a manufacturer calibrates its thermistors by measuring the three unknown parameters. A similar question applies to a user who receives the product sheet with the parameters or a lookup table but wants to validate them.

The calibration is done by subjecting the thermistor to a number of reference temperatures; in this case because we have three unknown parameters we need at least three known temperatures. The three reference temperatures and corresponding measured resistance provides three equations in three unknowns that can be solved easily. The reference temperatures can be achieved in a few different ways. One way is to use high-quality existing temperature sensors. The reference sensor and the one we are calibrating are both placed in a stirred fluid bath so that both sensors experience the same temperature. Another technique is to use physical properties of an ice bath or boiling water as reference temperatures.

The range of temperatures chosen for calibration and validation are chosen according to the expected range where the sensor is supposed to operate. For example, if the goal of the IoT sensor is to report back when a piece of equipment is too hot for humans to touch, the design can tolerate a significant amount of error in lower temperatures (say <70 °F) and do well above that.

Filtering the Sensor Output

Measuring physical quantities such as pressure, humidity, temperature, and others is always prone to noise and unwanted high-frequency jitters. There are natural high-frequency changes to any temperature measurement that our specific application may not be interested in tracking. Also, all the electrical components add noise to the actual signal. For all of these reasons, we need to filter the signal to get rid of the high-frequency components of the signal. If you are not familiar with the notion of a *filter*, think of it as a component that receives a signal as its input and attenuates certain frequencies in that signal before outputting the result. If, for example, the signal is audio, passing it through a low-pass filter means the filter lets the lower frequencies (low pitch, bass) through while attenuating the higher frequencies (high pitch, soprano). The same notion applies to any other type of signal. There are two common types of filters in the IoT industry. To get rid of high-frequency jitters in a measurement signal a low-pass filter is often used.

An analog filter receives an analog signal, filters it, and outputs the filtered signal as another analog signal. For less critical use cases, a simple RC circuit may be sufficient. However, if the output is fed into another circuit (display, analog to digital converter or ADC) that has low input impedance, it can result in an overdraw of electrical current. In other words, we will need to buffer the passive filter. In those cases, an active filter is often used which means leveraging some op-amps and a few more resistors and capacitors. It certainly adds to the cost of the whole design, which needs to be justified. However, the design of such filters are quite easy and commoditized.

Digital filters, on the other hand, act on digital signals. It means they need to be placed after the signal has passed through an ADC. Digital filters are made up of several flip flops, registers, and multipliers that hold the state of the circuit. Digital filters, like ADCs, need a clock on the circuit so that they can move from the current state to the next. It is practically important to have both the ADC and the filter run on the same clock. There are many commercial packages that take the desired cutoff frequency of the low-pass filter and generate a digital filter design. An important aspect of the digital filter design is its *order*. The order of a digital filter is a measure of its complexity and how many internal flip flops are required to implement it. As the order goes up, so do the degrees of freedom and the ability of the designer to customize the frequency response of the filter at higher frequencies. But like all other aspects of the design, higher order means more circuitry and more cost. Therefore, the designer should only care about the frequencies that matter to the use case.

Let's quickly touch on the cutoff frequency of the filter. In the example of a thermistor to sense the temperature, the thermistor itself comes with its own frequency response. In other words, the one factor to consider is how a thermistor responds to variations of the ambient temperature at different frequencies. If variations beyond a certain frequency get significantly attenuated by the thermistor itself, we may not need to worry about those frequencies. The manufacturers often provide this information in their product data sheet. The other important practical point is that the cutoff frequency needs to be set to much higher than the minimum theoretical numbers (Nyquist rate). That is because no filter is ideal; the actual attenuation starts at lower frequencies than the nominal one. While filtering unwanted portions of the signal, we are in fact changing the signal itself; therefore, we need to pay attention to the fidelity as a result of filtering. For simple signals and filters, a simple rule of thumb is to set the cutoff frequency to 5–10 times the highest frequency of the signal that we care about. The interested reader can read more about digital filters, and in particular Finite Impulse Response (FIR) filters, that are quite common in the industry.

FILE SYSTEMS

The gathered data and metadata are often stored in files on a drive on the device. The mechanism and protocols by which data is stored or retrieved is defined by the file system. If there is no file system in place, all of the data is stored in one huge object on the storage device without any type of optimization or ease of addressing done to it. It would be very hard to understand where an object starts and ends; writing and retrieval will be highly inefficient leading to system performance degradation. Other than optimization of writing and retrieval, file systems also provide structures such as directories, utilities to manipulate files, and custom management of rights and privileges to different users known as authorization and privileges. IoT systems are no exception in working with ever-changing data and metadata and, hence, are beneficiaries of the file systems.

While disks remain the most popular means of storing data, the file systems also help data storage on other devices such as flash drives, databases, and many more. There are even virtual file systems whose main purpose is to surface a physical file system in a different way as if the file system is something different. Some popular file system examples are FAT, NTFS, UDF, and HDFS. What makes a file in the file system is a series of blocks physically written in various and dispersed places in the storage device. The OS maintains a list of all of these locations and their order in a way that makes it efficient for data retrieval while traversing from one block to the next. Besides that, the OS also keeps track of file names, directory structure, and also each user's permission to each and every file.

One category of file systems act on blocks or sectors of a fixed size. In those systems, a block is defined as a fixed integer number of bytes that is the quantum of the number of bytes transferred. For example, the block size can be 512 or 4000 bytes. Every transfer of data will be in an integer number of blocks. For example, if an application asks to retrieve 700 bytes in a 512-byte block system, the operating system will retrieve 2 blocks. A look under the hood may also be useful. When an application asks for a number of bytes to be read from a file, the OS translates the request to the number of blocks and the file pointer (where the block begins). The OS then retrieves the data coming from the chain of blocks in its own buffer as an intermediary step. When the retrieval is complete from the file system, it then transfers the data from its buffer to the memory allocated to the requesting application. It is common for such commands to also return a status code to show success or error. Better applications handle errors in a detailed way to react accordingly in a safe and user-friendly manner. Another category of storage devices use objects instead of blocks. In these systems, there is a software layer sitting on top of the data that makes the handling of the data more flexible. The cost may be some more latency but in return applications can directly communicate with the storage device and also retrieve an arbitrary number of bytes. The retrieved data may or may not pass through an OS buffer.

Primary memory in embedded systems and computers refers to the memory that is directly accessible by the CPU. The registers inside a CPU or what we know as RAM are the main types of primary memory. These types of memory tend to be very fast relative to the other types of memory. They are also volatile, meaning that with the loss of power they lose their data. The secondary memory is the memory that the CPU cannot access directly. This memory is often used for slower and larger storage and access to data such as files for future use. Batch processing, which is performing operations on large amounts of data that is not time sensitive, is a common use case for analytics and reporting use cases. Generally speaking, computers have two orders of magnitude more secondary memory than primary memory. For example, when the RAM size is in the order of 1 or 2 tens of gigabytes on a personal computer, the same machines offer hard disks that are measured in terabytes. Hard disk drives (HDD) and solid-state drives (SSD) are the main types of secondary memory used in computers today. In many cases, these two are combined together for improved performance. Solid state drives are much more common for IoT devices. They use solid-state-based integrated circuitry to store data. This is in contrast with the traditional hard drives where an electromechanical system registers bits on a moving magnetic disk. Therefore, solid state drives run quietly and offer faster data retrieval. The technology inside these drives is often flash memory. Not surprisingly, the SSD is capable of providing much higher input/output operations per second (IOPS). A typical SSD setup can offer one order of magnitude higher throughput than a hard disk drive.

While the flash technology is mature and economical, it comes with its own downsides. One downside is data leakage. Even though the secondary memory is considered non-volatile memory, after several months flash memory can start leaking charges and losing data. The pace of data leakage goes up as the ambient temperature rises. Therefore, solid state drives cannot be kept without power for extended periods of time. Another drawback is that their driver software tends to be quite complicated. The reason is that physical limitations of the flash drives dictate that the same bits be rewritten in new physical places in the circuitry in case of changes to the stored data elsewhere. In order to make this regimen sustainable to the OS and outside applications, the software that drives the memory should keep track of all of the logic that maps the physical locations to the logical addresses at speed at all times. Nevertheless, these drives offer a fast, clean, and noiseless storage technology that's been quite popular.

Network File System (NFS) is a file system that was developed by Sun Microsystems in 1984. It is a distributed file system in that the actual files on the file server can be located remotely from the client machines. Interestingly, the file system was originally built by Sun Microsystems in a limited fashion for its internal engineering needs. Upon success and proven business value, it was turned into an open protocol and became commercialized (also interestingly Amazon Web Services was created originally for internal engineering needs of Amazon). A valuable aspect of the protocol is that the client and server systems are decoupled and can be running different operating systems and hardware systems altogether. For example, a Windows client can see a file as if it is in one of its own folders even though the file physically lives on a Linux file server. The actual drive can be a magnetic hard drive or SSD. While innovative and groundbreaking at the time, NFS is not widely used or suitable for big data use cases of today. It means transferring a few terabytes of data in a typical scenario can take hours. The main issue with the system is that it is built for one *scaled up* file server. While one server can be scaled up by adding more memory, disk, cooling, and power, there are serious limits to how much it can grow. More modern file systems allow for smaller servers to be joined in a *scaled out* manner and offer a much more elastic approach to file systems in a virtually limitless fashion.

The more modern systems that need to store and process larger amounts of data in the order of petabytes or more use parallel distributed file systems. In these distributed systems, data is stored in many smaller and potentially lower grade and cheaper machines. They also offer a significant improvement to the resiliency of the system by replicating data multiple times across the system and monitoring the health of each node. Many newer data storage systems, such as Snowflake, offer several more basic big data operations in-data-base so that massive data movement is minimized. For example, joining a few large tables with fairly complicated conditions followed by some basic statistical operations can be done without much macro data movement. Only more sophisticated operations such as building machine learning models with neural networks, or massive joins across two cloud systems would require more computational muscle that calls for the data to be moved. By moving as much of the computation closer to where the data sits, the system achieves two other goals in the meantime. It speeds up the execution of parallelizable algorithms significantly by avoiding the network latency caused by moving large chunks of data around. It also helps businesses save on the power consumption by

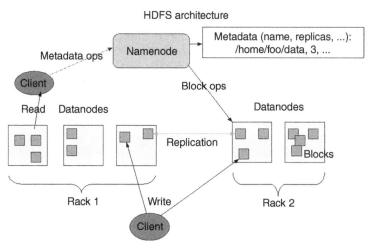

Figure 5.6 Hadoop architecture (*Source:* Image courtesy of The Apache Software Foundation).

saving the power needed to move the data from storage servers to the farther computation nodes. Here, we consider two popular distributed file systems: Hadoop and Lustre.

Hadoop Distributed File System (HDFS) was initially released in 2006 by Doug Cutting and Mike Cafarella (fun fact: *Hadoop* was Doug Cutting's daughter's toy elephant's name at the time. It also explains its elephant logo). HDFS first splits large files into blocks and spreads them across nodes in HDFS. Each block is replicated three times for fault tolerance. At the time of computation the code is shipped in packaged format to these nodes. As long as the desired algorithm is parallelizable, the primary (master) asks each node to perform the computation *locally* in a process that's called mapping. Once the results are ready, the server collects and aggregates the results in a process that's called reducing. The HDFS architecture is shown in Figure 5.6. The combined MapReduce process was originally quite obscure and difficult to work with in the original subsystems that made Hadoop. Since then other frameworks, such as Apache Spark, have made the lives of developers easier by encapsulating some of the intricacies of the original MapReduce process and bringing them to a higher level. The HDFS, however, remains quite popular. It is very common these days to have a big data system run against Apache Spark whose file system is HDFS. Hadoop's mission is stated as follows:

- Mitigate the impact of inevitable hardware failures
- Offer both streaming and batch processing of the data
- Provide support for large data files; the block size is 128 MB, while the theoretical maximum file size is the staggering 2^{89} bytes
- Offer write-once-read-many scheme for data integrity
- Offer support for heterogeneous and diverse hardware systems across different software systems

In the architecture diagram, the Namenode is in charge of opening and closing files when a client node needs to open or close or rename a file. The Datanodes keep blocks of the files and serve them on request. The Datanodes are responsible for serving read and write requests from the file system's clients. The Datanodes also perform block creation, deletion, and replication upon instruction from the Namenode. Note that the clients don't need to worry about the details of replication, addressing, or aggregation of the blocks that happen under the hood between the Namenode and Datanodes. For example, if one of the three copies of a block becomes unresponsive due to hardware or power failure on one of the nodes, the primary node detects that and automatically orders one of the available nodes to keep a new copy of the missing block. That way, the fault tolerance is preserved automatically.

The Lustre File System was intended to run on Linux. The name comes from combining the words Linux and Cluster. It was developed in 1999 and offered to the general public in 2003. Unlike Hadoop that was written in Java, Lustre is written in C. It is available under the GNU General Public License. Lustre provides high-performance file systems ranging in size from small clusters to large-scale, multi-site systems. The level of distributed prowess offered by Lustre is impressive. It can support clusters of tens of thousands of client nodes, scale its storage to petabytes of data (300 PB as of 2021) across hundreds of server nodes, and offer a staggering aggregate IO rate of more than a terabytes per second. Unlike HDFS, data replication is not a focal point of Lustre but could be accomplished as custom configuration. Similar to HDFS, there are Management Servers and Metadata Servers that the clients use to find the locations of files and handle other administrative affairs of the cluster. The Object Storage Targets, which can be commodity inexpensive machines, store and serve the actual data. One or a group of Object Storage Targets are managed by an Object Storage Server that routes the requests to the storage nodes. The server nodes that are in charge of Metadata, Management, and Object Storage Servers support the notion of failover and thereby make the operation of the cluster resilient to fault at the server level.

Data in Motion versus Data at Rest

Imagine a scenario where an oil rig drill is equipped with connected sensors to gauge the pressure at the tip of the drill. These drills can be quite expensive to fix, so you really don't want to exert too much pressure on them during the operation. On the other hand, after the operation is complete you want to go back in time, look at the collection of data from several rigs, do batch analyses, and find patterns and best practices and anomalies. These two scenarios point to the concepts of data in motion and data at rest, respectively. These real-world requirements represent an uncountable number of similar scenarios from almost all industries.

What we saw so far for data storage mainly handles data at rest. When the pressure at the tip of the drill is gathered and organized for several rigs, they are stored in a big file system for future modeling and analysis. But when the data is moving from the sensor toward the server, we call that data in motion. The architecture behind data in motion is quite different from data at rest. Modern software platforms and protocols such as Apache Kafka, Apache Pulsar, and RabbitMQ all concern themselves with handling large amounts of data in motion. While all three are open source, Apache Kafka is the more widely adopted platform by enterprises. When an entity in the network has a data stream that might be interesting to others, Kafka allows the entity to *publish* the stream. The interested entities can then *subscribe* to the stream to get updates when new data becomes available. This mechanism in Kafka is called pub/sub. While horizontal scaling is built into the design to support the growing number of streams, low latency is another major factor for time-sensitive use cases. One of the Kafka benchmarks (courtesy of Confluent) guarantees delivering 200K messages at under 5 ms latency 99% of the time.

Real-time operations such as alarms or actuation, as well as visual dashboards, are fed through real time subscription to streams. The same stream is bifurcated to also feed the longer-term storage of data at rest so that the future batch analysis can be done. This architecture is sometimes called the Lambda architecture (visually imagine the bifurcation in the letter λ). Large cloud providers, such as AWS, provide such architectures as a service so that IoT professionals don't need to start from scratch. But of course with any such choice, there is loss of control, as well as potential monetary cost ramifications.

MACHINE LEARNING

In this section, we review machine learning as a major driving force behind process automation and optimization not only in industries but also improvements in consumer quality of life. After a general review of the concepts that constitute machine learning and artificial intelligence (AI), we focus on different ways to leverage this technology for IoT applications.

History and Background

Since very early in human history of thinking, philosophers have been thinking about the meaning of learning and decision-making. In modern times, the most formal approach to artificial intelligence is attributed to Alan Turing. He devised the Turing Test that to date is sometimes used to gauge intelligence. The test proposes that a human and a machine perform a conversation through text. A third human observer monitors the text exchanges, and only text exchanges, between the two parties. While the observer knows that one of the participants is a human and another one is a computer she doesn't know which is which. If the computer handles the conversation in a way that the human observer cannot reliably determine which participant is a human then the computer is said to have passed the Turing Test. The thought behind the Turing Test is to propose a practical, simple, and reasonable way to define intelligence.

Since Alan Turing in the mid-twentieth century, the field saw another rise in hype and popularity in the 1980s. In those years computers significantly shrank in size and became ubiquitous in society. Computer scientists started to imagine a world where almost everything can be coded as a massive set of if-then rules that can be run by computers. The rise of the so-called expert systems was in that era where rule-based translators and expert physician systems popped up across the software industry. The idea was to encode all the human knowledge of those fields into logical statements; since computers are really good at following instructions, no matter how large, human-grade intelligence would be achieved. In other words, the attempted approach was quite heavy on logic and light on data. Soon enough everyone realized that the real world has way too many corner cases and rules to conform to a set of human-made rules. Translators started to spit out gibberish translations at a simple twist of the meaning and expert doctor systems easily confused words and symptoms frequently. With that failure, the field underwent another round of hiatus.

Starting in the new century, three main factors happened that rekindled the AI fire. First of all, data gathering and storage became significantly cheaper. That allowed systems to store and serve massive amounts of data efficiently and quickly at economical costs. Second, computer hardware such as faster CPUs, GPUs, and later TPUs enabled parallel processing of heavy algorithms in times that were orders of magnitude shorter than 10 or 20 years before. And third, based on the two factors above, computer scientists, statisticians, and engineers started to build new and advanced algorithms that *learned* how to make decisions based on a large number of past observations. Many algorithms boil down to a fundamental principle: based on the current state of the world and given all the historical observations, what prediction would minimize my likelihood of error. Here is a simple and intuitive example: if someone enters the room carrying a wet umbrella, we can predict several things about what has happened. Here are two: he could have poured water from a glass on his umbrella before coming in, or it may be raining outside. Based on most human's prior observations, the latter prediction is least likely to be erroneous. Of course, specialized algorithms can get quite complex. This approach is fundamentally different from that of the 1980s in that earlier attempts to build intelligence were mostly based on hard rules. The new mindset tries to mimic how humans learn most of their skills: observe the world around you, gather data, find patterns, and apply similar conclusions to future instances in a least error-prone manner. This approach has its own flaws. To name one, it typically cares less about the cause and effect in the real world; rather it concerns itself with correlations and best predictors. Another shortcoming is that the resulting models are as good as the data set that was used to train them. In other words, if the observations are poor or biased, the resulting model tends to be poor or biased as well. Ethics in artificial intelligence is a budding field in academia and industry. Having said that, we should add that the benefits of these modern algorithms, enabled by the advances in hardware and data gathering technologies, have outweighed their shortcomings in a significant way. The fact that a machine can effectively learn from past historical observations and make a specific decision has countless real-world applications of value.

Most existing algorithms are basically mappings from a large space of predictors to a space of predictions. In the most abstract way, a neural network that detects cats in images is nothing but a function from the space of all pixel values to a simple binary set of cat or not-cat predictions. The challenge is that the space of observations can be massively large and very hard to describe in traditional ways. This is fundamentally what tripped up the rule-based attempts of the 1980s. The modern techniques build a mapping so that a measure of error is minimized. Various models have their own measures of error and how to minimize them.

"All models are wrong, some are useful," said George Box, the famous British statistician who passed away in 2013. In other words, no one should expect a machine learning model to perfectly replicate the real world with its first principle rules. Our measurement and modeling capabilities are still insufficient for that matter. And that is not to mention physical limitations in measuring the real world dictated by theoretical bounds such as Heisenberg's uncertainty principle. We aim to create a mapping function from the space of observations to the space of decisions (prediction, classification, clustering, hypothesis testing, or others) that is *least bad*. If we can do that well many valuable applications will come out of it. For comparison, humans make erroneous decisions all the time. It is important to not compare machine learning with a perfect utopia but rather with the real alternatives in the world. And, at the risk of upsetting some philosophers, we should add that humans are essentially similar machines who experiment, observe, learn, and move forward.

Now let's translate these concepts to the IoT world. Imagine that a smart doorbell is supposed to identify motion and, therefore, trigger an alarm. Motion can come in many different shapes and forms many of which would not be of interest to the owners of the system. The owners would want to be notified only when a human being or an animal approaches their front door; they don't care if a car passes by. The machine learning algorithm needs to make a decision whenever there is an object moving in its frame. Can the decision be 100% correct? Of course not. The way these models are built are to feed them with many images of animals and humans that are labeled accordingly. From this point on, the problem becomes similar to a Bayesian probabilistic question: given all the observations in the past for which we already know the answer (labeled data) what is the likelihood of this new observation to be a human or animal? In typical models based on neural networks, certain features are extracted from the image such as ears, face, arms, fur, and others. The next layers of the model then hierarchically go up from there and connect different individual features to each other to form bigger features such as faces, bodies, torsos, and other larger objects. The algorithm then minimizes its prediction error by choosing the decision that minimizes a specific error function. That error rate or confidence is almost never 100%. However, as long as they are close enough, the users will take it. The definition of acceptable error rates is highly dependent upon the use case.

Another example is when a microphone next to an industrial pump is tasked with raising a flag when the pump needs service. Traditionally, this task is done on a fixed schedule or only when the pump stops working. When there is an experienced technician with *good ears* around, they can listen to the sound coming out of the pump and intuitively decide if the pump needs service or not. Needless to say, all of these approaches have flaws or are wasteful. Based on the frequency characteristics of many healthy and many faulty pumps' sounds a data scientist can turn this decision problem into one of machine learning: given all of the labeled healthy and faulty observations in the past how to label

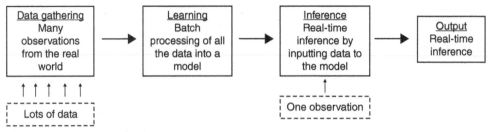

Figure 5.7 Machine learning schematic.

the new observation (the Fourier transform of the sound sample collected from the pump) such that the likelihood of error is minimized. Figure 5.7 shows the flow of data and different building blocks of the machine learning process.

While most of the recent advances in machine learning have been around these association-type inferences that is not all. Causality is a new and philosophically different direction where pioneers like Judea Pearl have been trying to teach machines to move from pure mapping to finding causes and effects within a set of observations. The old adage of correlation is not causation rings a bell here. In real-world scenarios, the next step after anticipating an event of significance is deciding what to do about it. This is where many professionals get tripped up. While certain variables or features can be great predictors of the output, they may or may not have a significant causal impact on the output. A simple yet real example is the correlation between customers attending webinars about the new feature in a product and them renewing their subscription to the product. While there might be a subset of the users who, by attending the webinar, become convinced to renew, most customer researchers believe that this is a fallacy. In reality, while attending webinars and renewing are quite correlated, the causal link is quite weak. In reality, there is often a third factor such as existing happiness with the product that drives both attending the webinar and renewing. Focusing on more webinars as a cause for more renewals not only can be ineffective, it can even frustrate some of the customers who are happy with the product but not interested in webinars.

Another real-world industrial example of this phenomenon happened when a team of data scientists from a renowned organization detected "the root cause" of an industrial equipment's failure only by observing the measured data. When they turned over their results to the subject-matter experts, they quickly discovered that the so-called root cause is nothing but a proxy for the alarm signal on the equipment itself. While this is an extreme example, such premature causal inferences happen very frequently in the industry. A new way of using observations is required to distill the cause and effect relationships. We encourage the interested reader to study the field in more detail in books like *The Book of Why: The New Science of Cause and Effect* by Judea Pearl and Dana Mackenzie. The one important takeaway here is to pay attention to the differences between good predictors and actual causes; they are related but not the same things.

As a decision maker or technologist, beware of the hype and fad around artificial intelligence as well. While these technologies are very valuable in automating and optimizing certain tasks, they are not magic. The industry appears to be coming out of the fad phase and grasping the reality of what this technology can do. However, there are still numerous examples where executives to subject-matter experts alike presume that a black box that is empowered by artificial intelligence can suddenly solve massive business problems with little effort. Such advances in the operations through AI typically takes a significant amount of effort by several teams in any organization. Later in this book, we will dive more deeply into the process of making a data product effectively.

Given the importance of the role that humans play in making machine learning work, we need to reiterate the fact here. In the processes of developing the right problem to solve, gathering and engineering the right data, and the interpretation of the results, humans' fingerprints are all over the process. This has nothing to do with the automation prowess that AI brings with itself after implementation. This is an important point not only to bust certain myths about AI power, but also to get ahead of some humans' fears of losing their jobs. AI in many cases helps humans become better decision makers or to focus on more interesting problems. While there is some truth to the fact that some jobs will be eliminated in the process, the positioning of the technology should be far from a pure replacement. In the author's experience, a large amount of the advances in the application of AI in consumer, and business use cases have led to greenfield and untapped use cases that have been infeasible otherwise. An example is alerting the owner of a camera doorbell whenever a human being approaches the door. In many other cases, humans benefit from the extra assistance they receive from AI; an example is when a set of connected and smart sensors alert a process owner that the state of a chemical process is anomalous and needs attention. In the following segment, we review some of the important concepts in machine learning and more common machine learning algorithms.

Categories of Learning Algorithms

In the most common AI and machine learning pieces in the literature, we run into the concept of *fitting a model* to our data set. A model is a mathematical representation of a large data set. For example, we can measure the current, i, and voltage, v, across a resistor one thousand times and record all one thousand pairs of voltage and current readings in a data set. This is the raw data set. Once we decide to represent the behavior of the resistor through a linear equation of the form $v = R.i$ (for a constant positive number R), we have represented this data set through a linear model. In this simple example, we already knew the shape of the relationship between the two variables through physical first principles. In the machine learning literature, oftentimes such first principle prior knowledge about the relationship does not exist. On top of that, the number of variables is usually much bigger than our toy example above. Besides, the relationship between the variables can be highly nonlinear. Consider the following example. We need to build a model that describes the power consumption of an office building through several readings of the power consumption and a list of other variables. These variables are usually called *features* in the language of machine learning. The features we record can be the following long list of disparate continuous, discrete, or categorical quantities:

- Number of people in the building
- Outside temperature
- Day of the week
- Time of the day
- Direction of the wind
- A measure of cloudiness
- Whether or not the AC is on
- Whether or not the heater is on
- Each server rack's on/off status

Needless to say, this is a much more complex model to try and build compared to our simple resistor example. Building a mathematical relationship between all of the above features on the one hand and the power consumption on the other is far more difficult due to the number of features, their various types (continuous and categorical), and the nonlinear nature of their relationship. Unlike the simple resistance example, typical machine learning algorithms can come in quite handy here; in contrast, in the resistor example, anything more than a simple linear regression will be an overkill. There are various types of modeling techniques in the world today. Many of them have been commoditized and packaged in commercial or open source offerings that can easily be used by experts and non-experts alike. Different machine learning algorithms possess various pros and cons that make them suitable with one type of problem while unsuitable for another.

Linear regression is one of the most basic and familiar ways of relating predicting features to the output of interest. Each feature will be multiplied by a constant factor and the results are added together. By training the model we find the constant coefficients that, when predicting the outcome on a test data set, minimize a measure of prediction error. While this model is simple to fathom and explain, it falls short if the relationship between input and output features is nonlinear. For example, predicting if an image, represented by many pixels, is a human face or not is way too complex for a linear equation to describe. That's why more complex models such as decision trees, random forests, K-nearest neighbors, and neural networks become important. Typically, more complexity brings less explainability and higher cost and power consumption with it. As a rule of thumb, *the least complex model that does the job is the best choice*. The important takeaway here is to be aware of such compromises when choosing an algorithm for a specific problem. In the next chapter where we talk about building a data product, we will review some of the trade-offs when we pick a machine learning model.

While we focused on predictive algorithms, there are many other types of models that are of high value in both consumer and industrial applications. Clustering is one of those categories where the goal is to cluster a large number of observations into an integer number of groups where each group is composed of similar observations. An example is clustering images of wafers in a semiconductor fab to classes that are visually similar for further evaluation and failure analysis.

Another class of algorithms are classification models. While clustering is categorically unsupervised (meaning there is no prior training involved), classification algorithms usually take a sufficient number of labeled data points for the algorithm to get trained. The training process can be quite time consuming and costly if the algorithm of choice is more complex. For example, a neural network that detects animal images may need tens or hundreds of thousands of animal and non-animal images that are labeled accordingly before it gets trained sufficiently. This is yet another reason to stick with the simplest model that gets the job done as more complex models tend to call for more labeled observations. Some researchers and startups are tackling this problem to enable model training with significantly less data points. Their approaches range from information theoretical to statistical techniques that are beyond the scope of our discussion here.

The next class of algorithms deal with text and speech and are generally referred to as Natural Language Processing or NLP algorithms. In the context of IoT such algorithms are often used in voice-controlled use cases such as smart speakers. Other use cases include sentiment analysis, word cloud, automating the process of finding answers to questions, and topic modeling. Typically, these algorithms first *tokenize* the words into words that are known to the algorithm and then push them through a sophisticated neural network. The types of neural networks that are capable of translating from a language to another or continue a conversation within the *context* can get quite complicated.

Interested readers should consult with specific text books on AI. The field has grown massively in the twenty-first century. There are several other areas of machine learning and AI that fall outside of our scope in this book. Reinforcement learning, generative algorithms, and several other niche use cases of AI are exciting topics for the interested reader.

General Machine Learning Concepts

The first concept to consider when fitting a model to a data set is the notion of prediction error. Everything else being equal, the smaller the error the better. However, it is extremely important to know that error is only one of several aspects to consider when choosing one model over another. In reality, it is quite possible to pick a model with higher error compared to other existing models due to some other crucial factors. That is in the same way that we don't always pick the most convenient sofa because it may be too expensive, require specialized cleaning, doesn't fit our designated space, its color does not fit the room, or a myriad of other reasons. Let's take a deeper look into how error is calculated.

If a model predicts a continuous number, such as our power consumption example above, the error is often defined to be the sum of the squares of the prediction errors for a *validation set*. In other words, only a subset of the available data set is used to build the model; it can be 70% of it for example. The remaining 30% is left untouched for the validation. When the model is built it is used to make predictions for every single observation in the remaining 30% of the data set. Each prediction result is then compared with the actual value. The difference is the error. To compare two models based on error, the squares of all of these errors are added together for each model separately. This type of error is often called R^2; its average over the validation samples is called mean squared error and is a very common way of measuring error.

In the case of classification models, say, to decide if an image is an animal or not, the error captures both false positives and false negatives. As an extreme case, imagine that our model always outputs the decision "animal" regardless of the input image. Such a model never incurs a false negative error by construction because it never predicts "not animal" to begin with. However, such a camera would inundate its owner with false alarms. This model is *sensitive* to an animal image because it catches them all; however, it is not *specific* at all because it labels everything as an animal. On the other end of the spectrum, a model that constantly classifies an image as "not animal" misses all real animals and hence is not sensitive at all, even though it never incurs a false positive error. In reality, classifier models are neither of these two extremes. There is, however, a compromise between the rate of false positives and false negatives. The choice of the model and its parameters is made in a way that hits the sweet spot between sensitivity and specificity. The actual choice depends on the application. The cost of a false positive for a life-threatening disease test is quite different from that of a doorbell camera that is supposed to detect an animal. The designer of a product should be aware of such compromises and pick the right combination for the real world application at hand. While the concepts may sound abstract in the beginning, they actually have significant real-world ramifications for a model. To gain a better understanding of false positives, false negatives, sensitivity, and specificity, see Figure 5.8.

How a Machine Learning Model Is Built

So far we have established an idea about how typical machine learning algorithms leverage historical observations to make a representative model. Now let's take a brief look at how these models are built. An abstract way of looking at these models is to imagine them as a function that maps an observation bundle to a prediction. In our building power consumption example, we considered nine features to predict the power. In that case, a model will be a function from the 9-dimensional space of predictor features to the 1-dimensional space of a power consumptions (a real number). In the classifier example, assuming that an image is made of a finite number of pixels, a model is from the space of all possible combinations of pixels to the binary set of {0, 1}. When a data scientist decides about the type of model, they basically choose the shape of the mapping function. For example, she may decide on a linear regression model that takes the shape of a linear function. The parameters of the function are initially unknown. In our linear regression example, the coefficients of the linear function are the decision factors. The learning algorithms' role is to pick these free parameters in a way that minimizes the error function. If the chosen error is the mean of individual errors squared, the algorithm picks the coefficients such that over all of the observations in the learning data set, the error is minimized. In mathematical terms:

$$\text{Model Parameters} = \text{argmin}\big(\text{error}\big) \text{over the training data set}$$

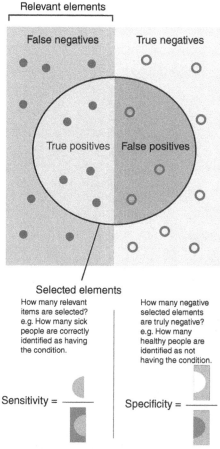

Figure 5.8 True and false positives and negatives along with the depiction of sensitivity and specificity of a classifier (*Source:* https://en.wikipedia.org/wiki/Sensitivity_and_specificity#).

So far so good! We are far from done though. One immediate question is, aside from simple linear functions, minimizing the error function over all possible combinations of Model Parameters can be a daunting task. This becomes even more prohibitive when we consider that certain modern neural network models can have thousands or millions of such parameters. Trying all possible permutations by brute force can take an extremely long time to complete that is beyond our lives. That is where the power of new algorithms combined with parallelized and powerful computing machinery (such as GPUs and TPUs) come into play. Major cloud operators and commercial machine learning platforms offer optimized algorithms and on-demand hardware to build models fast. Many of these algorithms iteratively move in the direction of reducing the error function until the consecutive improvements in the error function falls below a certain threshold. At that point, the algorithm is deemed converged to the optimal point. This family of

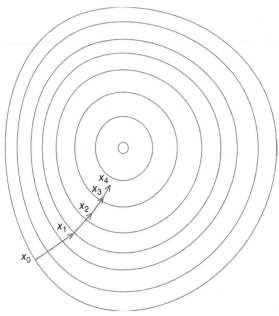

Figure 5.9 Depiction of the error reduction iterations through a gradient descent algorithm (*Source:* https://en.wikipedia.org/wiki/Gradient_descent#).

optimization algorithms are called *gradient descent*. In Figure 5.9, each contour curve is a level set of the error function. The gradient descent algorithm jumps from one level set to another iteratively until the improvement becomes small enough that it is deemed converged; think of it as gradually descending toward the bottom of a valley surrounded by hills. The speed at which the algorithm converges depends on the size of each of the iterative jumps toward the optimal point; the size of iterative jumps is determined by the algorithm *hyperparameters*. Model Parameters eventually determine the model itself and the properties of the model that represents the data set. Hyperparameters, in contrast, determine how fast the learning happens. Once the learning is complete, all we store is the model and its parameters to perform predictions. Choosing the right hyperparameters is a highly nontrivial task. Too big jumps can result in unstable algorithms that oscillate around the optimal point or diverge (escape) altogether. Too small hyperparameters can result in learning processes that take way too long to converge and, in the process, cost lots of electricity and money.

Overfitting

The astute reader may ask "What keeps the model from fitting all the points in the training data set exactly?" Imagine that we have two points on a plane. We can always find a line that goes through both points *exactly*; in other words the error will be zero.

We can pass through these two points an infinite number of higher degree polynomials such as three dimensional planes. In all of these cases, the mean squared error is exactly zero. We can pass a three-dimensional plane and infinitely more higher-order planes through any given three points. By extension, for any finite number of data points, we can increase the degree of a polynomial to the point that the resulting function passes through all of the points with zero error resulting in a *perfect fit*. "What's wrong with that?", you may ask. That's where the notions of overfitting and underfitting come in.

When we are trying to learn the behavior of a very complex system, we cannot do that after making too few observations. For example, no reasonable person expects to be able to build a model that describes the power consumption of a six-story office building by only measuring the power for an hour. Trying to do that would be a case of underfitting, which is a cause of high error in prediction. A less obvious cousin of underfitting is overfitting. As noted above, we can always exactly describe the whole learning data set with zero error by increasing the complexity of the learning function. That is also bad because of a phenomenon called overfitting. Note that what we need to learn is the general behavior of a physical system as opposed to just the finite observed samples. Each observation is only one observation at a specific point in time. Back to our extreme two-point example on a plane, an infinite number of functions could have created those two points on the plane. By making the models unnecessarily complex, we increase the chances of introducing unnatural complexity in the system. Not only that, any measurement in the real world is the actual quantity plus noise. By insisting on learning every single bit of the information we are in fact forcing the model to learn the data and the noise together. The meandering curve in Figure 5.10 fits every single point while the actual physical system may very well be a simple linear function, which is much more probable in the real world. This is a true example of a case where higher error is preferred to zero error.

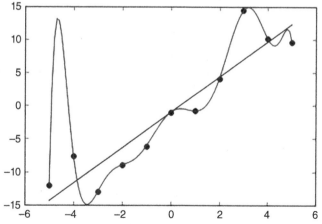

Figure 5.10 A simple linear function avoids overfitting the data while the meandering complex function overfits (*Source:* https://en.wikipedia.org/wiki/Overfitting#).

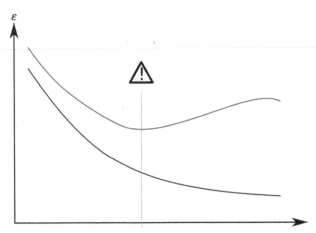

Figure 5.11 The error on the *y* axis is shown against the iterative training of the model *x* axis. While the training error continues to drop (bottom curve) the validation error starts rising at a certain point which indicates the start of overfitting (bottom curve) (*Source:* https://en.wikipedia.org/wiki/Overfitting#).

To address the problem of overfitting the data set is always divided into at least two or sometimes three partitions. These subsets are kept separately. The first set is used to train the model while the second set is not touched. The second validation portion is only used to test and validate different models. If a model overfits the training data set, its error rate on the validation set tends to be high (bad). That is because the model has been made too much of a good fit to the training data set by following any anomalous or noisy sample along the way. Such a strategy will show itself when the model is applied to the not-seen-yet validation data set. In Figure 5.11, the vertical axis represents the error, while the horizontal axis represents the iterations of the learning algorithm. The two curves show the progression of errors as the algorithm keeps iterating. The bottom curve (lower error) shows the error on the training set to constantly fall as the algorithm progresses. The top curve, though, shows the error incurred on the data points in the validation set. As expected, this error keeps falling to a certain point before it starts moving in the wrong direction. At that point, the data scientist knows that the model has started overfitting the training set. That's the point where the iterations stop. The main takeaway here is that mere minimization of the error on the training set is not the sole goal.

Data Set, Explainability, and Supervised Concepts

If the learning algorithm is the engine, data is the fuel. Data quality and quantity both are critical factors in the success of the resulting model. Good quality data means observations that are free from noise and are also labeled correctly. Such a dataset represents the population of interest adequately well without leaving significant subsets out. To continue our example of predicting the power consumption of the office building, note that if temperature sensors are noisy or not calibrated correctly, the resulting model will likely be low quality or completely off as well. In many algorithms, we need observations to be labeled correctly at scale. For example, if a security camera is supposed to identify humans as opposed to animals, we will need observations that are appropriately labeled as humans and animals as a starting point. Such algorithms typically need lots of labeled samples in order to base their learning. This part is sometimes a significant source of delay and cost to projects. There are crowd-sourced and commercial services that offer data labeling by humans for a nominal fee.

The quantity of the training data is also a very important factor. Different algorithms can have vastly different needs when it comes to training data volume. While simpler linear regression or decision tree models can produce decent results by hundreds or even several tens of samples, more sophisticated neural network models that perform image or speech recognition may require hundreds of thousands to millions of samples. You always need to consider the following few factors as trade-off decisions. In a vacuum, everyone will prefer a model that makes predictions with higher precision. However, improving the quality will come at a cost. In this context, we only consider the cost of training data, although there are a few different factors that we will review later. If the use case would not benefit too much from reducing the error rate from 5 to 4%, spending time and resources to acquire better and more samples may not be justified. Put more succinctly, the cost of improving the data set always needs to be compared with the perceived benefit of doing so. Making that judgment is not always easy. Business developers and product managers are often good at translating error rates to perceived dollar amounts so that the corresponding teams can compare apples with apples when deciding about extra data sets or compute resources. Experienced data scientists have a good intuition about how much data would be needed to meet certain levels of accuracy for various types of models.

In certain other machine learning applications, confidence in the result is an important factor that needs to be communicated to the stakeholders. For example, a product manager may demand that "a process manager needs to be notified immediately of the issue if the algorithm deems it to be imminent only at 95% confidence or higher." The reason behind such a request is to avoid false positives. Such false positives will inundate the operators with erroneous errors and, over time, will erode the confidence in the model as a whole. Another common trade-off in statistical models is between the sample size and confidence interval: generally the bigger the sample size the higher the confidence in the statistical inference. The same trade-off needs to happen here as well. Investment in improving the size of the data set needs to happen only when there is a clear business need for tightening up the confidence in the results. If the product owner is happy with 97% confidence in the results, spending money and delaying the project to make it 98% needs to be carefully reevaluated. Besides that, the incremental cost of model improvement increases as the model gets better. In other words, the cost of going from 50 to 51% is usually much lower than going from 97 to 98%.

For the more technically oriented reader, it may be helpful to know of a statistical technique called bootstrapping. This technique is sometimes used to improve the quality of prediction or a hypothesis testing inference without having to increase the size of the data set. In essence, the technique randomly resamples from the existing data set many times with replacement. The technique was developed in 1979 when data collection was costly and difficult in many situations.

Explainability is a sometimes-overlooked characteristic of a machine learning model. So far what we discussed was all about the quality of the model prediction and its cost. Explainability concerns itself with the transparency of the underlying model and its decision process. Some models are called black box models because they are practically impossible to explain. Take our process error system example again. It's one thing to say that "the machine learning model identified an alarm situation that was correct and the process manager was notified." It is a different offering to say all of the above plus "and the model says that because the temperature of the second furnace has remained over 300 degrees for more than 5 minutes." That is a very critical point for certain use cases. Some models, however great at making inferences or predictions, are mathematically built in a way that makes them extremely difficult to explain. Winning stakeholders' trust in those cases will be harder. Besides, certain industries, such as pharmaceuticals and financial services, are highly regulated. Embedding models in their processes to make decisions need to be backed by a clear and comprehensible explanation of how the model makes its decisions. Generally speaking, neural network models can be quite difficult to explain even though they can be very effective as a predictive model. In contrast, a decision tree or linear regression model can be explained in plain human language much more effectively even if they lose some level of prediction accuracy. This is another important factor to keep in mind when designing machine learning systems.

Most of the examples mentioned so far are examples of *supervised learning*. In this class of models, there is a labeled data set that is used to train a model. The model then makes decisions for new and unknown observations based on the learning. An example is a security camera that is designed to identify a particular face; if the face is a preset familiar person, no alarm is raised. Otherwise, a notification is sent to the authorized users. In this example, the training data set is the image of the familiar person's face that is labeled as "the face of interest." The model extracts facial features, such as eye color, shape of the smile and lips, and other features to train its internal model. When facing any face, it uses the pretrained model to decide if it belongs to the familiar person or not. Another example is when an industrial system is being trained to alert operators when a piece of equipment is at risk of failure. In this case, sensor data, as well as metadata, belonging to similar pieces of equipment are fed to a machine learning model. All such data sets are labeled as either normal or pre-failure. The model then learns how to map the data to a normal condition or a pre-failure condition. In this case many observations may be needed to train the model adequately. The exact volume of the observations depends on the algorithm chosen and the desired accuracy. Simpler algorithms, such as decision trees or linear regression models, tend to deliver reasonably good results with hundreds or even tens of data points. Neural networks and deep neural networks need many more observations before they can learn the data.

The opposite of supervised learning is *unsupervised learning*. In this case, the model does not take labeled observations to learn from. For example, a wind turbine operator may be interested in learning how many *modes of operation* can describe the behavior of a turbine over time. You can think of this intuitively as clustering various types of behavior in the equipment based on measured data. The result can then be used to identify abnormal situations, gradual drifts, or failure in measurement sensors as the root cause. In this scenario, observation data is fed to a training model. The unsupervised model splits the data into a finite number of clusters. Typical algorithms allow the data scientist or operator to adjust the number of clusters to their liking. Aside from that they do not need to provide the model with labeled data. During runtime when a new observation comes in, it gets assigned to the closest cluster. Any deviation from known and normal clusters can be labeled as anomalous.

What Is and What Is Not Machine Learning?

So far we reviewed the general concepts behind machine learning and some common terminologies. We also saw a few categories of models and corresponding techniques. Now let's take a step back and talk about the problems that machine learning is good or not so good at solving. Machine learning techniques are very good at finding patterns. They can quickly absorb a large amount of data and find similarities, anomalies, patterns, or clusters in them. After the learning phase, which typically requires a significant amount of data, they can make real-time decisions quickly. These capabilities make them very good at image recognition, anomaly detection in sensor data, pattern recognition for failure or risk prediction, and forecasting events based on prior observations.

There are a few important assumptions that need to be satisfied for these capabilities to realize. One of them is the existence of a sufficient amount of data. What is sufficient, you may ask. That highly depends on the chosen algorithm, as well as the desired accuracy in the model. For example, a simple power consumption forecasting model that relies on a dozen features and is based on a random forest may provide sufficiently good results with a couple of hundreds of observations. But if the goal is to detect a human passing by a security camera at 98% or higher accuracy, the model typically needs at least several tens of thousands of labeled images to get trained. Another important point is that not all observations are created equally. In a predictive maintenance use case, the model needs actual pre-failure data to be able to train itself on the conditions that indicate an imminent failure; providing the model with a huge data set that is all from normal working condition may not be enough. More generally, the amount of information may vary significantly between two pieces of observation in the training dataset. To see this point, imagine that we are interested in learning a person's dietary habits. If we have observed their breakfast for two months already, adding the data about another breakfast may not add much information. However, if we observe their lunch for the first time, we can gain much more information. By the same token, if there are specific portions of the space of observations that are already sampled enough, we may not gain much by adding more of the same to the training data set. The reverse is true in that by adding observations from sparsely represented segments of the observation space, we can help the model improve significantly. There are statistical techniques in machine learning that adjust a dataset so that the training is not skewed too much in favor of over-represented parts of the population. If these adjustments are not done the resulting model will be prone to bias.

Another assumption is that the recipients of the model are willing to let a machine learning model make a decision for them easily. This is less of a technical issue and more of a human and process aspect. We will discuss this topic in more detail later in the book when we cover the general process of building a data product. It suffices to say that the technical capability of a machine learning model is not a guarantee that it will be accepted

Figure 5.12 The human brain is very good at building the causal chain of events after making observations. Current machine learning algorithms are not so good at that. (*Source:* https://en.wikipedia.org/wiki/Hammer_throw#).

and implemented in the real world for various reasons. Prior study of the culture and the processes and regulations impacted by the model needs to happen before a model is built.

Now let's turn our attention to the types of problems for which machine learning, as we know it today, would not be a good fit. The astute reader may have noticed that nowhere in machine learning capabilities we mentioned reasoning or causation. This is probably the biggest shortcoming of existing algorithms today that still distinguishes a human brain from artificial intelligence. For example, a human brain can learn by itself that when a human throws a hammer in an athletic event (Figure 5.12), it is the human body that is the source of the motion to the cord that in turn moves the hammer. However intuitive it may sound to you, the existing algorithms are bad at making such causal connections. There are some efforts to make such causal inferences from observed data; however, the results are still considered at research level and applications are quite limited. It is important to note that the old adage of *correlation is not causation* holds true and strong here. It is a dangerous trap that many data professionals and subject-matter experts find themselves trapped in when building a predictive model. Consider this simple example: if I give you the shoe size of a person and ask you to predict if that person can read and write, you will find the shoe size a great predictor. Typically, a shoe size of six or higher (US shoe sizes) means the person is six years old or older, which is a great indicator of that person's literacy. While the shoe size makes a great predictor, you do not expect to buy a toddler a bigger shoe and expect them to be able to read and write. In this case, there is a third factor, age, that correlates heavily with both the shoe size and literacy and causes both of them. In real-world scenarios of machine learning applications, people sometimes mistakenly take the predictors with high impact on the prediction and take them as causal levers. Always be aware of this distinction and apply subject-matter expertise or controlled experiments (such as A/B testing) to draw causal connections.

Machine Learning Architecture

The architecture of machine learning for IoT applications follows the general IoT architectural layers: cloud, edge, and fog. We explain these architectural elements in more detail in the general IoT architecture section of this book. It is worth, however, to specifically review the choices and their use cases when it comes to machine learning in particular.

Machine learning in the cloud assumes that the data from the sensors all travel to a central location where a cluster of servers are gathered. The cluster or clusters are what often people refer to as the cloud. With that cluster architecture comes heavy computational power, economy of scale, centrally managed services, and cheap storage at scale. Many use cases lend themselves to a full cloud architecture. For example, when a wind turbine operator wants to look at the data from 1000 turbines over a year and do aggregate analysis of turbine performance a cloud environment is an attractive choice. The sheer volume of the data and the required computational power make the cloud an inevitable choice. Also, all the data sets from the turbines need to be aggregated in one place for the models to be built. Again, a central entity with sufficient privileges that is secured is required.

There are downsides to the cloud architecture. One of them is that all of the data needs to be transferred over the network. The bandwidth and its associated cost can prove to be prohibitive in some use cases. Latency is another issue. In the same example of wind turbines above, imagine that the goal is for the wind turbine to immediately stop or change the angle of its blades to protect itself in the case of a detected wind gust. Forcing the data to go all the way to the cloud for the machine learning decision to be made before the command comes back to the turbine can waste several precious seconds. Another potential issue with the cloud approach is privacy. Certain use cases, such as personal health data gathered through wearable devices, are subject to governmental regulations and consumer sensitivity. Requiring every bit of the consumer data to traverse to a central cloud can violate certain regulations and turn off more cautious consumers. Regulations such as GDPR codify some of these regulations that restrict data movement.

Before moving on from the cloud architecture let us make an important distinction. The learning phase of machine learning requires much more data and compute power. It is also the part that is not time sensitive. Most of the learning (or batch processing) happens in the cloud or on beefy local servers. The part that can be time sensitive is often the run-time inference phase of machine learning. Our examples above highlighted this distinction naturally.

Machine learning can also be done at the edge of the network. Ideally, a tiny battery-powered IoT device is made smart enough that it makes the inference right when the data is generated. The benefits of this approach include lower latency, lower power consumption, decreased security risk due to less data movement, less bandwidth requirements, and fewer privacy concerns. However, some serious challenges need to be resolved before this architecture can work in a practical machine learning application.

The main challenge of machine learning at the edge is power and memory constraints. To give you an idea, the size of a fairly simple neural network that is listening to detect a word in audio signal or is watching to detect a specific object can easily be more than 1 MB. Bigger models are more power hungry; they can require hundreds of milliWatts or even more. In many IoT applications, none of these is an option. Over the past few years, there have been many efforts to do more with less when it comes to trained models. Custom-building a neural network instead of taking a massive cookie-cutter model to train is one of the major ideas. The tinyML foundation was founded in 2019 with the goal of making machine learning models so small that they fit small form factors of IoT devices. Certain successes have been reported such as shrinking a model that listens to a wake up phrase to just over 10 KB. Shrinking trained models to 100 KB and less is the foundation's general goal. Some people in the community are hopeful that even the learning (batch processing) can someday be done frugally at the edge; we are not there yet.

Machine learning in the fog layer is, in certain ways, the middle ground between cloud and edge. Fog is the set of compute and memory resources that lie between the edge and the cloud; or the cloud that is closer to the user and where the data is generated (hence the name *fog*). Many industrial use cases have aggregator nodes on-premises that gather and join data from various IoT devices together before sending to public or private clouds. These nodes are connected to the wall power and are often significantly more powerful than the frugal edge devices. Many modern smartphones can also be considered powerful computers with much less-stringent power restrictions; therefore, they can serve as aggregators and perform fairly heavy machine learning operations without sending the data to the cloud. Many machine learning use cases can be performed in the fog layer. As the architecture implies, the use cases and benefits are somewhat the middle ground between cloud and edge. Some bandwidth saving is achieved in the backhaul where data moves from fog to the cloud; but the bits still need to move from the edge to gateways and fog servers. Similarly privacy, security, and latency benefits of fog architecture fall somewhere between edge and cloud architectures. The specific requirements of each use case and allocated budgets determine the criteria for the optimal choice of architecture. The next segment of this chapter goes more deeply into the topic of IoT architecture.

NETWORKING AND COMMUNICATION SYSTEM, PRODUCT LIFECYCLE, DEVICE MANAGEMENT

In this section, we go deeper into various aspects of an IoT system. First, we study the three main architectural layers based on the logical distance from where the data actually gets generated. Then we turn our attention to connectivity technologies and architectural considerations for various networking topologies. In that process, we also review some of the more common physical communication technologies with their proper use cases. We conclude the segment by device management and longer-term lifecycle analysis of the IoT system.

IoT System Architecture: Edge, Cloud, and Fog

There is a common way of looking at the IoT system architecture based on the logical distance from where the data is generated or, sometimes equivalently, where the user interacts with the system. For example, imagine a series of sensors spread around an oil refinery. They measure all sorts of physical and chemical quantities right where the physical phenomenon happens. This is called the *edge* of the IoT system. The edge of the system is the closest you can get to the source of the data. It is where the data is at its rawest and most voluminous shape. Also, the edge of the network is typically the most resource-constrained part of the network. These sensors and devices can be running on small and inexpensive electronics, be geographically dispersed, have limited battery power, and otherwise be in a rugged situation. The primary role of the edge is to interact with the source of the data: most of the time collect data and sometimes convey commands in the opposite direction. Because of these common limitations, the compute expectations from this layer is limited. Although, there may be a paradigm shift happening as more efficient algorithms tend to push computation closer to the edge. Having said that, we should add that simple rule-based logic and threshold comparisons can be run very easily and cheaply on small pieces of hardware. For example, when the temperature exceeds a threshold, an alarm is raised or a signal is sent to the gateway. It is a common technique to send data up to the cloud only when certain conditions are met that imply the data is worth processing in a special way; that way bandwidth and power are saved while events of interest still are reported back to the central system. Things can get more complicated if we expect the small sensor to run statistical process control.

The power and hardware limitations of the edge have led the industry to form clusters of computers that are connected to an endless source of power for heavier calculations. These clusters are called the *cloud*. Cloud systems are managed by centralized administrators and can scale up or down based on the transitory ebbs and flows of the compute requirements. That way cloud systems can not only maintain responsiveness in their tasks, they can also be economical by ramping down their compute power when there is no demand present in the network. Cloud systems can enjoy the economies of scale due to their aggregated nature. Many different end users and systems can coexist in one cloud and be separated logically for privacy and security reasons. Complicated cloud management systems allow system administrators to operate massive clouds efficiently. Today, there are a few frontrunners in the market followed by several more niche cloud operators that have their strengths in special vertical markets.

Another applicable scenario for the cloud systems is when the analysis of interest requires aggregated batch data from various edge devices and possibly other systems over a period of time. For example, imagine a smart watch manufacturer who uses two brands of chips in their smart watches for tracking the users' heartbeat. After a while it needs to see if there is any difference between the quality of one chip provider and the other one. To answer that question the manufacturer needs to gather data from many edge devices, enough of each group, and combine (join) that with its own internal manufacturing data tables before analysts can draw statistical conclusions. Aggregating that combination of required data can only happen in a cloud environment.

Now the natural question is, if the cloud systems offer so much more power, why don't we always use the cloud instead of the edge. There are various reasons for that. The first reason is latency. If every bit of decision needs to be made at the logically and often geographically remote cloud system, the decisions can face delays. Every hop in the network adds a nominal delay, which is exacerbated by the roundtrip nature of the decision process. On top of that, adding any extra hop adds a risk of failure and unexpected issues, which can lead to complete failures. If the use case of interest is time sensitive, making the decision closer to the edge is preferred. Some of the smart railway systems are good examples of this requirement where delayed or failed decisions can be catastrophic.

Another factor that favors edge computation over cloud is the cost of moving the data. Imagine an extreme scenario where edge sensors are merely conveyors of raw data all the way up to the cloud; no single decision is made by them. Therefore, every single bit of measurement needs to be pushed through the network. Such a broad brush approach can be quite costly in more than one way. First of all, depending on the connectivity technology, bandwidth can be costly. This is especially true in case of remote sensors that are connected through satellite or cellular networks. Another issue is power consumption of battery-powered sensors. Certain communication protocols, such as Wi-Fi, are quite power intensive. Forcing edge devices to send lots of data can drain their batteries very quickly. There are also security and privacy considerations. Adding any new link or physical node that data passes through adds yet another attack surface to the security of the whole system. The data that never leaves a sensor deep on the refinery floor will be virtually impossible to be hacked by a remote hacker. Privacy becomes a concern especially for consumer IoT systems. Many users are wary of their security camera footage being uploaded to a big vendor's cloud; if face recognition and other machine learning algorithms run on the camera itself, the consumer will feel more comfortable.

Therefore, it's important to know where to use each system and how they complement each other. Edge computation is very good at responding quickly, avoiding unnecessary bandwidth cost, preserving energy, and enhancing data hygiene. On the other hand, cloud systems are very good at economically performing massive data operations, merging and joining various data sources including non-IoT ones, performing batch analysis and building machine learning models, and otherwise being the heavier and more powerful engine in the whole system (Figure 5.13).

In recent years, there has been a movement to enable smaller edge devices to do more operations where the data is collected in the first place. The fact that security breaches in large ecosystems have been on the rise makes the edge increasingly attractive. On the one hand, hardware advances have made tiny and inexpensive systems quite powerful. A simple observation of how personal smartphones have evolved over the past 20 years proves the point. From the other side of the problem, researchers have been quite creative in shrinking the size of heavy algorithms so that they can run economically on small form factors. For example, take the use case of a smart speaker that is constantly listening for its wakeup phrase (such as "OK Google") followed by the actual command from the user.

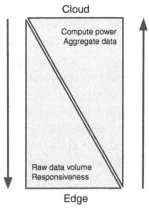

Figure 5.13 Edge cloud diagram.

The algorithm that detects the wakeup phrase from the audio signal has to run at all time. Researchers at Google have shrunk this algorithm to just over 10 KB so that it can run on a small chipset at mW levels of power. Only when the wakeup call is detected, the bigger processes wake up to process the user's more complicated command. While we expect the miniaturization trend to continue to make progress, we also expect that cloud and edge will continue complementing each other in a synergistic way.

Now it's time to touch on the concept of Fog Computing. Cloud systems offer massive storage and compute capabilities. Wouldn't it be good to bring some of those capabilities closer to the edge of the network? That sounds like a cloud that comes closer to the ground, or Fog. In order to organize the disparate efforts toward this goal, the OpenFog Consortium was founded by ARM, Cisco, Dell, Intel, Microsoft, and Princeton University in November 2015. Since their inception, the consortium merged with the Industrial Internet Consortium (IIC). Based on their reference architecture document, the Fog System is defined as "A horizontal, system-level architecture that distributes computing, storage, control and networking functions closer to the users along a cloud-to-thing continuum." In other words, this architecture defines all the required pillars, such as communication, security, and others, so that more functionality can be accomplished without having to push the data to the cloud. Interested readers can review the publicly available reference architectures by IIC and the Open Fog Consortium.

Networking

Connecting all the components of the IoT system together in a way that meets the use case criteria while it keeps the data safe and secure is a nontrivial aspect of the IoT system design. There are many different sides to this challenge. In this segment, we look at various terminologies, architectural choices, and physical communication protocols.

The high-level nature of making the communication happen can be done in a few different ways. One of the older approaches, that still has certain use cases, is called *Store and Forward* networking. In this approach, the data is stored on an intermediary node, checked for health and integrity, and then forwarded to the (next) destination. Some of the email servers operate in this way in case the connectivity is not reliable. In IoT scenarios, this scheme can be useful when latency is not a concern. This approach is good for when delays can be tolerated without significant consequences. When sensors are embedded in very remote locations, such as sensors for longer-term climate studies, the gathered data can be transferred altogether at a later time. At the same time, submitting the data usually includes a handshake to the next destination which consumes power. Given the remote nature of these sensors, it is often more economical for the sensor to do the handshake fewer times and transmit the data in batches to save on precious battery energy. Some such sensors communicate through satellite communication services that may not offer 24/7 connectivity depending on the location of the orbiting satellite; hence, being able to store and forward is necessary.

When latency is critical, *cut-through switching* is an option. In this type of networking, a switch or intermediary node starts forwarding the data (packets) as soon as it knows where it is headed. That instant can happen before the whole packet is completely received. That way the latency is reduced. Obviously, the outgoing channel needs to be at least as fast as the rate of incoming data. A price to pay for this lower latency is loss of ability to check the packets for error. While in the Store and Forward approach, the packets are checked for integrity on the forwarding node before submitting, cut-through switching is not able to do that. The reason is that by prioritizing speed, the switch is instructed to start moving the data before it is completely received. Checking for integrity is left to the receiver. This approach is suited for delay-sensitive applications such as sensitive alarms or real-time audio/video communication systems. Also, if the nature of the network is not so much prone to faulty packets, the overhead of passing occasional faulty packets may be negligible.

Deep packet inspection or DPI offers a more modern and thorough approach to packet switching in a network. In simpler switching approaches, only the necessary headers of a packet are inspected to determine the destination and routing information for the packet. Therefore, if an intrusion has caused an anomalous application to flush information out to a suspicious destination, the routing system will not be able to catch it. DPI, on the other hand, inspects much more deeply inside the packet to glean many more aspects of the communication by inspecting the data portion of the packet on top of the routing headers. Security is a major use case for DPI. Another use case for enterprises is to enforce policies depending on the type of the data being transmitted. On the flip side, some privacy advocacy groups express concern about such algorithms that allow deep inspection of otherwise private data. In fact some governments have used this mechanism heavily to do surveillance on the types of data being transmitted or to block certain applications altogether.

TCP/IP

The packets in the network consist of actual data, as well as the metadata that conveys information about the source and destination and other aspects of the information within the packet. In order for various computers and networks to communicate with each other, a popular protocol was developed in the 1960s by DARPA that is called TCP/IP. The name is short for Transmission Control Protocol/Internet Protocol. This protocol defines all the aspects of how packets should be formed, transmitted, routed, and received. In what follows, we review the four abstract layers that are defined by TCP/IP. Understanding the concepts behind these layers is crucial in understanding how communication between machines happens. In a more granular model, called OSI Layer model, the protocol is divided into seven distinct layers. However, there is a clear correspondence between those seven layers of OSI and those four layers of TCP/IP. In the interest of simplicity, we will only focus on TCP/IP here. The layers and correspondence between the two models are shown in Table 5.1.

The Network Access Layer or Link Layer is where the rules and protocols corresponding to the physical transmission of packets between two nodes in the same network live. The protocols typically live in the driver software or firmware of the devices. The protocols are hardware agnostic enabling broad applications. Link Layer eventually delivers the packets to the physical medium for the data to traverse to the next destination.

The Internet Layer is responsible for the logical rules and protocols of transmitting packets over the entire network; this is in contrast with the Link Layer that is responsible for intra-network movement of packets. The Internet Layer is where the routing in the larger network happens. There are three main sub-layers within. The Internet Protocol or IP Layer looks up the destination based on its IP address in the packets. IPv4 was the dominant protocol for many years before IPv6 emerged as the newer model of addressing to handle the shortage of available addresses. The emergence of many IoT devices was a big factor in the shortage. The Internet Control Message Protocol or ICMP allows the hosts to communicate potential problems in the network. The third subcomponent is called Address Resolution Protocol or ARP whose job is to map physical addresses of hardware such as MAC addresses.

As you can imagine moving packets through all of these protocols reliably can be a massive overhead for any application. The Transport or Host-to-Host Layer wraps several of the details in the lower layers for the applications to not have to worry about most of the details under the hood. In contrast, the User Datagram Protocol or UDP provides a simple

Table 5.1 Seven layers of the OSI model compared with the four layers of the TCP/IP model.

OSI Layer model	TCP/IP Layer model
Application Layer	
Presentation Layer	Application Layer
Session Layer	
Transport Layer	Transport Layer
Network Layer	Internet Layer
Data Link Layer	Network Access Layer
Physical Layer	

way of sending packets over a network without too much overhead. It does not offer any measure of correctness or reliability. On the other hand, Transmission Control Protocol or TCP offers acknowledgment of packets for error-free communication. This extra feature naturally comes with a significant amount of overhead in the number of headers. If reliability is not a major concern UDP may be sufficient, while TCP is for more error-sensitive applications.

The highest layer in the TCP/IP Layer model is the Application Layer. This layer leverages all of the services of lower-level layers, and especially the Transport Layer, to offer user services such as peer-to-peer or client-server communication. Example protocols in this model are the familiar HTTP, HTTPS, FTP, or SSH. HTTP is short for Hypertext Transfer Protocol and is used by the web browsers to connect to servers in remote locations. HTTPS combines HTTP with the security protocol in Secure Socket Layer (that we reviewed in Chapter 4, Security) to provide a high-level secure channel. Similarly, FTP is used for file transfer, and SSH offers a secure way to offer remote shell or terminal emulation.

Network Functions Virtualization and Software Defined Networks

The problem of optimizing network setup and maintenance in business and industrial environments is quite different from that in a consumer product. In consumer products, the number of connected devices ranges from a few to a few tens in the majority of cases. Moreover, the audience is not necessarily professional or technically savvy. Devices should be easy to set up and configure by nontechnical users. Usually, these devices are ready to connect to a short range radio out of the box, such as Bluetooth, so that they can be configured through personal phones. From there they continue on that communication medium at steady state (for example a smart watch) or configure themselves to connect to the residential Wi-Fi (for example a smart plug.) In contrast, industrial use cases can be much more complicated. The number of connected devices can be many thousands and the environment can be rugged with spotty or sparse connectivity. On the other hand, it is more reasonable to assume that the installers and maintenance crew are technical professionals. In what follows, we look at two concepts that are in particular valuable in the industrial IoT networks.

Network functions virtualization (NFV) and software-defined networks (SDN) are related and complementary technologies that help industrial users save on hardware and maintenance of their networks. NFV deals with virtualization of networking functions by abstracting them from the underlying hardware in order to save on hardware cost. Having every single one of the networking functions, such as switches and routers and firewall servers, on a separate box is costly and inefficient. This is similar to why having a few virtual machines on one piece of hardware saves money for IT teams in other contexts. One way to go about this problem is to use a blade server and have the networking applications run side by side on virtual machines or containers. NFV takes this idea one step further. On top of bare metal, a Hypervisor offers a flexible set of virtual machines that are logically separate. These virtual machines run side by side on one set of hardware. Not only does this technology save money it also offers security by keeping the virtual machines as separated as physically separate boxes. NFV also allows for prioritization of applications depending on their sensitivity to latency. In other words, if resources are short, priority is given to the routers and switches that deal with more latency-sensitive traffic. NFV combines forwarding devices and middleboxes into a common control framework. Network operators, as a result, can implement network policies much more easily due to the fact that many details of network modules are abstracted out. For example, the important concepts of placement of switches and steering of traffic are abstracted out and can be optimized through NFV. The former refers to *where* each function needs to be placed in the network and the latter addresses how the *forwarding* should happen.

Network operators can steer traffic based on the location of middleboxes or conversely move the middleboxes where there is heavier traffic. This optimization and routing can be quite complex. NVF helps solve this problem by intelligently managing virtual network functions and applying network policies. The policies, in turn, determine what network elements should run at what time and where in the network in a dynamic manner. In contrast, without NFV each one of these changes would need spare hardware blades in various locations of the network leading to significant cost increase. The policies also determine how to react to callbacks from each element in the network to provide fault tolerance and reliability. The infrastructure orchestration and management layer of the NFV also performs continuous discovery of the resources and traffic before allocating or removing elements in the network. Routing processes are determined by the resulting policy that is set forth by the management layer. The bottomline is that NFV's goals are to minimize the footprint of implementing the policy and to enable optimization for bandwidth, latency, and network resources much more easily. Flexibility of NFVs allows network administrators to run a more secure system by treating each stream of traffic based on its unique source, destination, or corresponding priority.

An SDN, in turn, divides the network functions into a data plane and a control plane. It also provides the mechanism to dynamically determine how network traffic flows through a network. While SDN and NFV are related, they are not the same. You can have NFV in place without SDN, or the other way around. NFV abstracts the policy and forwarding function from the hardware that they use to run. SDN, in contrast, separates the network control function (control plane) from network forwarding functions (data plane). In most cases, though, the two concepts are used as complementary to each other. As an example, specific software-defined policies that are part of SDN can be implemented in a virtualized way enabled by NFV.

The control plane manages the network's performance and faults. It does that through protocols such as NetFlow, IPFIX, or SNMP. The controller knows all the details of the network topology and determines resources accordingly. The data plane or infrastructure layer routes the actual data using guidelines and commands from the controller. The switches along the way route the traffic based on the policy that is dictated by the control layer in a software-defined manner. SDNs allow lighter switches and routers to only keep track of more common policies; in case of corner or less common decisions, they can ask the controller what to do. That way, the switches can remain light weight. The application layer is then able to make high-level requests that are handled through the control and data layers.

Through this mechanism, the controller can minimize cybersecurity or scheduling risks by adjusting policies on the fly when new threats are discovered. In case of growing networks, SDN allows for scaling of the network through breaking up the network into multiple control and data planes for better modularity and control. Another way of scaling is to let local switches make most of the routing decisions locally. Some of the benefits of SDN are:

- Switching and routing can be done through inexpensive switches because of the extra dynamic flexibility that SDN offers.

- Optimization of traffic flows can happen on the fly.

- Automated logical topology of the network can be optimized over time especially if it is combined with NFV.

- Dynamic and agile response to routing needs is feasible.

- SDNs can filter traffic on the fly based on content and firewall rules through deep packet inspection.

- Since software changes are less costly than hardware swaps, SDNs help avoid vendor lock.

- SDNs make network administration easy by providing one central place to configure a network.

- For more complex commercial networks SDNs enable dynamic pricing.

A typical scenario that can be handled efficiently by an SDN is when the network requirements vary at different times of the day or week. When businesses are active, the control layer can allocate the majority of the bandwidth to business users during the day. That scenario shifts in the evening when consumers are back home and need more bandwidth. Overnight and when both businesses and consumers are less active, batch and large one-off jobs can be prioritized over the network. This way the same infrastructure can be flexibly optimized per the needs of the time.

WIRELESS COMMUNICATION PROTOCOLS

A strong majority of IoT systems include a wireless communication element embedded in them. In consumer applications users are accustomed to be free of wires for several years now. Depending on the power supply (battery or wall powered), the communication happens through a relatively close gateway or goes directly to the Internet. An example of the former is a smart watch that has to be very frugal about its battery power. The data that it exchanges often goes through the connected smartphone over a low-power protocol such as Bluetooth. On the other hand, a smart thermostat that is powered through wall wires can afford to directly connect to the Internet through Wi-Fi, which consumes much more power. In return for consuming more power, the network removes one hop (smartphone in this example), which creates dependency and latency. These trade-offs are quite common when choosing the suitable protocol.

In an industrial setup, when sensors can be powered through unlimited wired electricity, they afford to send more data at higher rates over longer range. For example, the emerging 5G protocol offers impressive latency and bandwidth specifications. In essence, sensors can exchange data at very high rates with extremely low latency at the cost of high power consumption. This scenario is very different from sensors embedded around water pipes in remote areas to detect leaks. In those cases, sensors need to be placed in areas where battery is the only choice and maintenance is costly. Therefore, they have to communicate only a few bytes once in a while only to meet the business specifications of the use case and preserve their battery energy. The communication can happen through long-range aggregators or satellite communication in more remote use cases.

In industrial use cases, there are sometimes other factors, beyond power, behind choosing short-range communication to a gateway before connecting to the Internet. One aspect is mobility. It is a powerful concept to be able to measure anything at any desired part of a manufacturing process. In the past, such changes would require significant wiring and electrical circuitry to be added; any such changes would be costly and require several levels of compliance approvals. Wireless connection opens the door to more granular customization of the sensor network and a more flexible manufacturing structure. Another aspect is security. Connecting small and less-sophisticated IoT devices directly to the larger network (Internet) exposes them to more risk. These are typically small and inexpensive devices that cannot afford to run heavy security protocols on them to defend themselves. Many organizations choose to leave some more beefy gateways in the middle to act as a layer of protection. Those gateway devices can be secured and managed by IT organizations efficiently.

In the remainder of this section, we review multiple wireless communication protocols in the market today. We mention their strengths and weaknesses and review typical use cases. We divide these standards into two main short-range and long-range categories. The short-range technologies can cover up to about 100 m, while the long-range ones can go as far as a satellite can fly.

Short-Range Wireless Protocols

The short-range standards in this section cover ranges as small as 4 cm up to about 100 m. Their use cases are for sensing, point to point connection, secure on-site configuration, personal area networks, and local area networks. A very familiar example in this category is when we listen to a podcast on our Bluetooth headset or connect to the Wi-Fi in the office.

Near-Field Communication or **NFC** or ISO/IEC 18000-3 has the shortest range on our list. Theoretical ranges are up to 20 cm, while practical ranges are often less than 4 or 5 cm. Historically, this technology has roots in the RF Identification or RFID where the goal is to keep track of assets from a short distance. NFC is a low-power and slow technology; while the tag can be passive (no power), the bit rate is up to 424 kbits/s. It operates in the 13.56 MHz range which, in comparison, is much lower than Wi-Fi or

Bluetooth that operate in the GHz range. As you can imagine from the original use case one of the two sides of the link is considered the reader and the other one is considered the tag. A very important characteristic is that the tag does not need power. When the reader's magnetic field gets close to the passive tag, the resulting inductive power causes the tag's antenna to transmit data. The data can then be read by the reader's antenna. This fact alone reduces the cost of NFC tags to a few cents apiece.

Security is an interesting aspect of this technology. Because of its low profile, the transferred data cannot be encrypted. However, due to its very short-range nature, the risk is very low to begin with. In other words, physical proximity to the tag is the only way where data can be read.

Let's look at some of the NFC use cases. Reading tags from a short distance for inventory purposes is one of the typical examples. Another example is for short-range file transfer between two devices. Most modern cell phones have the NFC reader and tag technology embedded in them. When two phones get close enough to each other, they can communicate through NFC and transfer files at a fairly slow rate. Therefore, this is typically suitable for communicating contacts or small pictures. Another use case is when more powerful communication (such as Wi-Fi) needs to be configured on a device in a protected environment. Instead of handing over the network password to other devices designers keep the network details in an NFC tag. That way, other devices can read the network information through NFC when and only when they are in physical proximity of the tag. Once they read the network information, they can get onto the larger network; this process is sometimes called bootstrapping.

Bluetooth is a household name and one of the most common technologies in the consumer IoT market. It has several characteristics that make it suitable for device configuration and personal area networks. The technology was originally developed in the 1990s by Ericsson Mobile. The idea back then was to replace RS232 cables with a wireless alternative that moves files to printers. The name Bluetooth comes from an ancient Danish King, Herald Bluetooth, who was able to unite many disparate Danish tribes. The idea is that Bluetooth similarly unites many different wireless technologies under one single protocol. Indeed the Bluetooth standard of today includes many different profiles and modes, which makes it quite versatile. Its use cases range peer-to-peer, as well as mesh networking. It works in the 2.4 GHz band with a range of up to 10 m in standard mode. In that mode, the protocol is ultralow power and consumes only 1 mW. In other classes of the protocol, devices can consume up to 100 mW and increase the range to 100 m. The practical range is highly dependent on the obstacles between the sender and receiver. The throughput is between 1 and 3 Mbps, which makes it suitable for voice and sound use cases.

Typical use cases of Bluetooth are small-to-medium file transfer, sound and music streaming, and device configuration. It is quite customary that a consumer IoT device arrives with both Bluetooth and Wi-Fi radio chipsets in it. At the configuration phase, the device uses its Bluetooth radio to be discovered by the nearby cell phone. It then exchanges its basic configuration data, such as its identity and Wi-Fi network and password, so that it can be configured and connected to the main network. Bluetooth offers various modes such as discovery of the device; this is when you make your headset discoverable by your phone. The protocol stack also offers audio streaming. Security is embedded in the protocol. The later versions of the protocol allow for AES security to be set up between the two sides of the link.

Bluetooth Low Energy or BLE is nothing but a special revision of Bluetooth that came out in 2011 and was officially called Bluetooth 4.0. This was a major enhancement to the protocol where power consumption was significantly reduced through the design of sleep mode when there is no need to transmit. That enhancement played a massive role in proliferation of the standard in various IoT devices that are power constrained. The technology keeps improving; in 2020, the Bluetooth Special Interest Group introduced BLE Audio that enables one device to share audio with multiple devices at the same time.

Zigbee or IEEE 802.15.4 was conceived in 1998, standardized in 2003, and revised in 2006. The name refers to the waggle dance of honey bees after their return to the beehive. It is a short range, low rate, low-energy communication protocol designed for IoT-type use cases in mind. The very low power usage combined with smart management of radio components can result in battery life of years. The communication is done digitally in packets and 128-bit AES security is part of the protocol. Therefore, the protocol is secure natively. On top of these aspects, there are a few other properties of Zigbee that make it especially attractive for IoT networks. The short-range nature of each individual communication link makes it suitable for smaller personal area networks; however, the mesh capability enables the protocol to serve larger area networks.

The protocol allows for self-forming and self-healing mesh networks. It means that first of all the devices do not need to be individually in the reach range of the gateway or base device; the Zigbee-enabled devices automatically form a mesh network to route packets from each node to the base device and vice versa. Besides that, if a node dies the network automatically reroutes itself. When a new node shows up, the other nodes discover it and make it part of the newly formed mesh network. The protocol is also built in a way that supports over-the-air upgrade to the protocol; that way if a new version of Zigbee is released, the nodes can be upgraded over the air without a need for staff to upgrade them manually.

The standard operates in the 2.4 GHz frequency range. Its range can go up to 100m depending on the obstacles. Its throughput is up to 250 Kbps, which means it is suitable for low bandwidth use cases such as sensor networks. The number of nodes in the network can theoretically grow to the massive number of 65000.

Its use cases range from medical devices across a medical facility to smart buildings and smart metering applications. Smart homes are a particularly good use case for Zigbee due to its simplicity and long battery life. For example, many smart plugs and light bulbs connect to their base device (that in turn connects to the Internet) through Zigbee; therefore, the sensors can operate on battery and last for years. At the same time, farther sensors in the corner of the yard can connect themselves back to the base through the mesh topology. The mesh topology can also be helpful in rugged industrial environments where signal reliability is low; having redundant paths can come handy in those situations. Reducing the need for specialized signal repeaters or range extenders helps lower the cost of deployment.

Z-Wave in some ways is a cousin of Zigbee. It is a low-power mesh protocol designed mainly to handle smart home use cases. We mainly focus on some of the differences between Z-Wave and Zigbee. Many principles of operation are shared between the two.

The first significant difference is that Z-Wave is predominantly designed and is present in smart homes; in contrast, Zigbee can be found in many commercial and industrial use cases as well. Zigbee is an open standard, while Z-Wave is closed and is controlled by Silicon Labs. Zigbee was adopted and was a central piece of Matter (formerly Project Connected Home over IP or CHIP) as we saw earlier in the book; notably Z-Wave was left out. This might over time lead to a significant boost in adoption of Zigbee over Z-Wave. However, Z-Wave has a better track record of interoperability among certified devices. In terms of technical specifications, there are a few significant differences between the two. While Z-Wave operates in the 908 MHz frequency band Zigbee is in the 2.4 GHz range; this might cause some interference between Zigbee and Wi-Fi but it is often not significant. Zigbee offers longer range (100 vs. 10 m for Z-Wave) and higher data rate (250 vs. 100 Kbps for Z-Wave). The maximum size of the network is another point of differentiation. Since Z-Wave is mainly built for smart homes, the number of nodes is capped at 232 while Zigbee can support up to 65000 devices in the network.

Wi-Fi is probably the most common and known brand of the communication protocols. The official standard is the family of IEEE 802.11 that was initially released in 1999. Wi-Fi was then coined as a catchy brand name, which has been quite successful. Since then several generations of the standard have been released successfully to enhance the rate, range, and security of the protocol while maintaining an impressive level of backward compatibility. Since IEEE itself did not want to get in the business of certifying individual vendors or devices, Wi-Fi Alliance (Figure 5.14) was formed in 1999 to bridge that gap. In 2019 alone more than three billion devices were shipped globally with Wi-Fi capability on them. The protocol is categorized under short range communication networks. It offers significantly higher data rates while consuming more power than Bluetooth or Zigbee.

Figure 5.14 Wi-Fi alliance was formed in 1999 to certify individual vendors and devices (*Source:* https://en.wikipedia.org/wiki/Wi-Fi#).

These facts make the technology suitable for video, image, and very high-quality sound communication. Smaller form factors, such as smart watches, may struggle keeping the Wi-Fi on at all times because their battery gets drained quickly. In those scenarios, the IoT device needs to be connected to wall power (such as smart plugs) or enjoy a sizable battery (such as smartphones).

There are different modes of operation in a Wi-Fi network. The most common mode is the Infrastructure mode where every device in the network connects to a single base station; this is what most of us are used to in our homes and offices. There is also an ad hoc Wi-Fi mode where enabled devices can connect to each other directly without a central base station.

The standard operates in two common bands of 2.4 and 5 GHz. Each frequency band offers several channels of communication. Historically, each upgrade of the standard has improved upon its throughput, security, and range, among other things. The more recent releases of the standard have been Wi-Fi 4 (802.11 n – 2008), W-Fi 5 (802.11 ac – 2014), and finally Wi-Fi 6 (802.11 ax – 2019). The next standard Wi-Fi 7 (802.11 be) is under development.

The range offered by the newer generations extends to several tens of meters up to 150 m in some cases. This number can be drastically reduced in case of indoor use cases where walls and other metal objects exist. The throughput rate of Wi-Fi 6 has officially reached 9.6 Gbps, which is at least an order of magnitude larger than those of Bluetooth and Zigbee. The standard offers security measures to protect against intrusion. Since any base station advertises itself with a beacon signal, any willing device can listen and try to join the network. Wired Equivalent Privacy or WEP was introduced originally to secure the communication. Since then and after several vulnerabilities were found, Wi-Fi Protected Access, or WPA, was introduced by Wi-Fi Alliance. Over time improved versions of WPA have been released, the latest of which is WPA3. It offers wider choices for password selection, as well as AES encryption of the data. It is generally recommended to not limit the security of the network to Wi-Fi security only. Besides hardening the wireless network as much as possible, the users should leverage VPN and HTTPS in both commercial and residential use cases.

Another specialty about Wi-Fi for IoT is the Wi-Fi Protected Setup that allows for small devices without a user interface to be set up on the Wi-Fi network. Adding Wi-Fi capability to an IoT device can be accomplished easily these days through many small and inexpensive electronic modules that are commercially available.

ANT and **ANT+** are the other protocols in the short range communication family. They are mostly used in health, fitness, sports, and medical devices. ANT is a proven protocol and silicon solution for ultralow power (ULP) practical wireless networking applications. It supports star, peer-to-peer, and mesh topologies in the 2.4 GHz frequency band. Its range is less than 10 m and offers rates of 1–2 Mbps. The protocol enjoys a vibrant developer ecosystem and offers a rich set of tools to license and build devices with silicon all the way up to Android and iOS environments. The first protocol (officially nRF24AP1) was introduced in 2005. ANT and ANT+ are managed by ANT Wireless, a division of Garmin Canada Inc. The ANT+ Alliance is an open, special interest group of companies who have adopted the ANT+ promise of interoperability.

Given its ultralow power profile, ANT can run on a small coin cell battery for several years. Its use of memory is also optimized so that it can fit small form factor devices; the capability can easily run on a single chip. At the same time, the protocol is smart enough to self-adapt when changes happen in the mesh network. Overall, the protocol is designed for small and ultralow power devices that should run for a long time.

ANT+ wraps the basic ANT protocols and offers definitions for interoperability among various devices. For example, a blood sugar sensor that communicates to its base through ANT+ uses basic ANT and publicizes its specific profile as a blood sugar monitor through ANT+. Other similarly branded ANT+ devices can then communicate to this device speaking the same language.

EnOcean is another ultralow power wireless communication technology. Unlike ANT, however, its use cases are focused on smart homes, spaces, building management, and some industrial use cases. The nonprofit EnOcean Alliance was formed in 2008 to foster innovation and growth of the wireless standard across the globe. The frequency band varies in different parts of the world to meet local regulations but all of them are sub 1GHz. A popular feature of the technology is energy harvesting. The commercial EnOcean organization, based in Germany, has built several sensors that harvest energy from ambient temperature, movement, and light. The goal is to create a low-maintenance automation network that does not require battery management. Naturally, the corresponding wireless technology is ultralow power. The indoor range is about 30 m while the open-space range can extend to 300 m. The maximum bit rate is 125 Kbps. In order to achieve the extreme power efficiency, the standard relies on short telegram (down to 1 ms duration) of messages with as little overhead as possible. It supports 128-bit AES encryption.

WirelessHART (based on IEEE 802.15.4), owned by FieldComm Group, is mainly used in industrial automation use cases. It was built to improve the existing wired HART standard in order to bring all of the benefits of wireless communication to the industrial world. The original HART protocol enjoys a massive install base of several tens of millions of devices in the broader industry and manufacturing world. The technology offers native mesh networking technology for reliable communication in harsh and rugged industrial environments. While some control and instrumentation systems come equipped with WirelessHART feature in them, there are separate commercial WirelessHART adaptors in the market that can be attached to existing instruments; that way legacy instruments can connect or even possess smart features.

The communication is two-way and happens in the 2.4 GHz range. It supports 20–250 Kbps data rate, which is often sufficient for massive sensor networks in industrial environments. In a typical architecture, up to 100 nodes connect with one gateway that gathers the data and backhauls it to the main networking infrastructure of the organization. Security is also a big point of emphasis for the protocol. It supports 128-bit AES, as well as features like rotating keys, device authentication, and reports on failed connection attempts.

6LoWPAN, which is an acronym of IPv6 over Low-Power Wireless Personal Area Networks, is another low-power radio protocol for smart cities, smart homes, and industrial use cases. The protocol is in response to the need for having remote and tiny form factors, such as smart grid sensors, to be able to join IPv6 networks. It achieves low power by compressing the headers in the IPv6 communication protocol. Thread, which is now one of the main protocols inside the smart home interoperability standard Matter, is built based on 6LoWPAN. The technology supports mesh topology for farther reach and resiliency in rugged environments. The frequency band supports both sub GHz as well as using the 2.4 GHz band.

Long-Range Wireless Protocols

Sometimes there is a need or preference for sending the signal over longer distances than just a few hundred meters. Use cases cover remote areas where there is no economic justification for building a networking infrastructure; some examples are wildfire sensors, meteorological meters in remote areas, soil moisture sensors for agriculture, and gas pipe leak sensors in vast pipeline networks. In all of these scenarios, the sensor itself needs to communicate with the next point of connection over distances that can easily exceed a few hundred meters or even several kilometers. Another class of use cases are when the asset of interest is mobile. For example, tracking trucks, mining equipment, ships, or other types of mobile vessels all require connectivity in motion.

In these situations, power management becomes exceedingly important. On the one hand, long distance communication consumes more energy. That is in the same way that talking to someone who is far from us takes a good bit of shouting. On the other hand, in the majority of these scenarios replacing batteries for the sensors is difficult because they are in remote locations. A combination of smart radio protocols and frugal communication can make a big difference in energy consumption. To make the communication more frugal, system designers are very careful about what to send and avoid communicating a single bit of data unnecessarily. In some situations, embedding some basic levels of rules and analytics on the edge device can be very helpful so that the device can decide on its own what to send and what to filter out. Besides the power considerations, the designers should also factor the pure dollar cost of communication over cellular or satellite networks. Even though those technologies have reduced their prices continuously over the years, each is still considered significantly costlier than a common local area network on Wi-Fi.

Cellular communication is extremely popular in both consumer and business use cases. Many countries in the world enjoy a rate of penetration of more than 90%, which makes cellular communication quite appealing, feasible, and economical. Among the growing standards 4G LTE is the most widely adopted in most nations while 5G is up and coming.

LTE is short for Long Term Evolution and is a trademark owned by the European Telecommunications Standards Institute. It has a very rich and crowded history of development and adoption in the early twentieth century that is beyond the scope of our discussion here. The standard made cellular packet switched communication possible and popular for mobile devices. Even though the standard has specific designations for phones and other voice use cases, IoT devices can leverage the packet switched mode to communicate bits and bytes to cell towers. The bandwidth is asymmetrically divided which allocates up to 300 Mbps to download and 75 Mbps to upload. These rates are in ideal scenarios; obstacles, interference from reflecting objects, and distance from the base station are factors that limit the rate. The latency can be as low as 30–50 ms. The communication is between each node (sensor or IoT device) and a cell tower with support for mobility and handover between cells. Speeds of up to 310 miles/h are supported without loss of connectivity. Therefore, it is a suitable choice for trucks and ships. The service providers are typically local to the geographical markets. Many such providers have specific plans for data-only nodes that can connect through a cellular module on the board of connecting devices. Small modules for cellular communication are available commercially for IoT use cases that are ruggedized and will fall back on 3G communication should 4G LTE fails.

The emerging **5G technology** provides a significant boost to data rate and latency, among other things over 4G LTE. Commercial network providers started deployments in 2019 and services are already being offered in many countries. The standard operates in three possible frequency bands that are dubbed low, medium, and high modes. The most important takeaway is that there is a trade-off between range and data rate. The higher the frequency band, the higher the bit rate and lower the range. The high mode operates in the 24–47 GHz range. By doing so it offers data rates as high as 20 Gbps in ideal scenarios. However, its range is significantly shorter and sensitive to walls and other physical obstacles. As a result it requires more antennas to form smaller cells, which makes sense for more dense urban areas. The low-band 5G offers data rate and ranges comparable to 4G. Therefore, 5G is designed to cover various types of use cases under one umbrella standard.

When there is a need for extreme low latency, such as remote medical use cases, gaming, or self-driving automotive decisions, 5G offers a latency of less than 10 ms. The latency highly depends on the implementation details; also, latency during handover between two cells can be significantly higher than latency while connected to the same base station. The expectation is for this technology to continue to grow in adoption and popularity in the 2020s. To put things in perspective here is a quick timeline of the cellular technologies of previous generations:

- 1G was introduced in the 1980s and offered analog voice service with mobility. It enabled the very first cell phones.

- 2G in the early 1990s digitized voice communication for better quality and added text messaging.

- 3G in the early 2000s made data services possible for the first time in a limited fashion.

- 4G came into the scene in the 2010s and made mobile broadband possible.

- 5G is being deployed in the early 2020s and improves on 4G by an order of magnitude in both data rate and latency.

LoRaWAN specification targets connecting IoT devices over wide area networks that can operate on low power and battery-powered form factors. The name comes from smashing the words in *Long Range Wide Area Network* together. The technology is the most popular low-power WAN in the world for IoT devices across agriculture, smart cities, buildings, manufacturing, logistics, and utility verticals. The standard is owned by the LoRa Alliance, which is a nonprofit association. The association's mission is enablement, promotion, and development of the LoRaWAN technology in the world for IoT networks. They also have an active certification program that certifies devices to connect with LoRa networks. Per the LoRa Alliance's description "The LoRaWAN specification is a Low Power, Wide Area (LPWA) networking protocol designed to wirelessly connect battery operated 'things' to the internet in regional, national or global networks, and targets key Internet of Things (IoT) requirements such as bi-directional communication, end-to-end security, mobility and localization services." There are more than 100 public and private LoRaWAN network operators across more than 170 countries. For example, in the United States today, there are a handful of public and open network operators offering LoRaWAN services to IoT networks across the country.

The LoRa sensors communicate with a LoRa gateway that, in turn, backhauls the aggregated traffic. The standard offers end to end 128-bit AES encryption, as well as over-the-air software and firmware upgradability. The protocol is in particular resilient against physical obstacles and can offer ranges in the order of a few kilometers. While 5G offers similar targets for coverage, their power consumption and data rates are the main differentiation. LoRaWAN offers data rates that range between 0.3 and 50 Kbps, which is a few orders of magnitude smaller than that of 5G; in return, LoRaWAN battery operated devices can last for years on battery.

Ingenu is another low-power and long-range technology that is proprietary to the company with the same name. It is, therefore, in the same class as LoRaWAN. However, its adoption and coverage is smaller today than its bigger cousin. The technology is secure and emphasizes reliability quite heavily. There are Ingenu networks across a few tens of countries, some of which are privately built for internal use cases. One Ingenu tower can cover a range of 50 km in any one direction in ideal situations. As a real example, 17 towers cover the Dallas/Ft Worth area that spans 2000 sq. miles. The standard is better suited for smaller data use cases because of lack of high speed in the standard. The proprietary technology underneath Ingenu is Random Phase Multiple Access (RPMA), which is a wireless communication protocol designed and built for IoT and machine to machine communication. It operates in the 2.4 GHz spectrum band that is available in most countries. Battery-operated devices communicating over Ingenu can last for several years or even more than a decade. Similar to many of its competitors, it offers 128-bit AES security.

WiMAX is another broadband wireless technology that has a mixed history. The technology is based on the IEEE 802.16 specification, which has been around since the early 2000s. The name is short for Worldwide Interoperability for Microwave Access which is owned by WiMAX Forum. Its features have a lot of similarities with 4G LTE and some argue that in the 2010s, the mobile broadband market was already won by LTE. Nevertheless, the technology still exists in pockets of use cases that require mobile broadband access. Its range extends to a few miles, which is an order of magnitude longer than Wi-Fi. It supports mesh topology, which can increase its practical range significantly. A significant difference between WiMAX and Wi-Fi is that WiMAX supports mobility. It can operate in the 2–60 GHz range depending on local frequency regulations. The data rate can go up to 1 Gbps but is highly dependent on the distance. Typical use cases are broadband access in areas where installing cable or fiber is difficult; installing fiber in dense urban areas can cost as high as $300 per foot. Between Wi-Fi, 4G LTE, and 5G, some argue that WiMAX will have a difficult time growing significantly beyond its limited adoption today.

Satellite communication is a broad technology for global and remote connectivity. IoT has been one of the main drivers of satellite communication technology. The rule of thumb is that if connection and backhauling are doable through links on the ground at an economical rate, organizations choose on the ground options over satellite. However, that is not always feasible. For example, a remote fire detection sensor in the heart of a forest is too far from any established communication infrastructure. Another example is a large cargo ship that spends most of its working days far from the land and ground infrastructure. Power grid monitoring and remote environmental sensor systems are a few other examples.

There are two common topologies possible for supporting satellite IoT use cases depicted in Figure 5.15. *Satellite-IoT backhaul* topology is boosted by low-cost and low-power short-range standards to connect a large number of endpoints and IoT devices to a hub with technologies such as LoRaWAN. Such transmitters' cost is now under $5 per unit. The hub, which is not power constrained, then connects to the satellite as the backhaul to send and receive data and connect to the larger network. The *direct to satellite topology* is better fitted for highly dispersed sensors or devices where building a local network doesn't make sense. Obviously, satellite connectivity in these cases need to provide low-power options given that battery replacement for such remote devices can be costly.

The operational margins for satellite service providers are not very high these days. However, that may be changing thanks to the exponential growth of IoT devices on the one hand and investment in low orbit satellite communication on the other hand. Incumbent

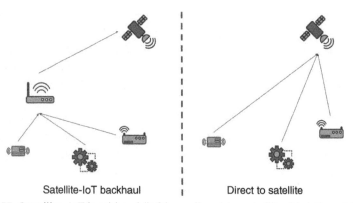

Satellite-IoT backhaul   Direct to satellite

Figure 5.15 Satellite-IoT backhaul (left) vs. direct to satellite (right) architecture. (*Source:* Web Vectors by Vecteezy).

companies including Thuraya, Inmarsat, Iridium, and others already serve a few million IoT devices across the globe. They are well equipped to serve as the backhaul service for many more IoT devices each of which may send infrequent small packets. Newer players in the market like Astrocast, Lacuna, Kinesis, Swarm technologies, and a few others offer low-orbit and low-latency satellite connections that can be suitable for direct connection to the satellite (as opposed to backhauling). SpaceX is another big name to mention in this market. Its constellation of tens of thousands of small and low-orbit satellites cover an increasing area across the globe that can serve residential use cases, as well as machine communication. As of winter 2021, they offer bit rates in the 50–150 Mbps with latencies in the 20–40 ms range. They are expected to continue improving the quality of their service as they launch more satellites in orbit.

The biggest game changer for satellite communication business has been IoT. The global market for IoT-focused satellite services, focused on end-device connectivity hardware and the connectivity fees, is forecasted to grow to $5.9 billion by 2025, after taking off in the 2021–2022 period.

EXAMPLES

Before we close this chapter, we briefly look at two examples, one from the consumer world and another one solving an industrial use case. The intention is to bring the concepts of the chapter together for better understanding. Needless to say, each use case can be the subject of a months-long project by professional teams. Our goal is not to design all of the parameters of the systems in detail, but rather offer a schematic and blueprint of the design.

Smart Video Doorbell

Let us review the high-level process of designing a smart video doorbell. Business developers do a market analysis about existing opportunities, total addressable market, and the competitive landscape. Once that process is done and the direction of the product is determined, the product management needs to do the ballpark of consumer research and feature design. In this example, let's assume that the product managers envision a doorbell system that enables the homeowner to answer the door from anywhere over the Internet. On top of that, they want to add video and audio communication, so that the two sides can see and talk to each other. Besides these basic functionalities the consumer will be enabled to leverage security and monitoring features with the always-present smart video camera in front of their door. Such a security system will inform the owner of any events of interest with as few false alarms as possible. Moreover, the user will be able to review their video recordings of their front door at a later time or share them with their neighbors. Members of the same family should be able to have equal access to the video feed and other main features. The setup and configuration process should be straightforward enough that a typical nontechnical consumer can easily set them up on their own. This is an important distinction in priorities compared with industrial use cases. While ease of configuration is always a positive factor, it is generally more important for the consumer market. All of the above features should be thoroughly documented in writing with sufficient illustrations.

The design and architecture teams have a number of high-level decisions to make before engineers start to build the software and hardware. To start they need to figure out the type of the camera that meets the resolution, wideness, and other requirements of the product managers. Also important is the form factor and power consumption of the camera so that it fits the required size of a doorbell. Most of these design decisions, while significant, are not too controversial. A potentially watershed decision is if the device should be wall powered or battery powered. This decision alone will impact the addressable market and several engineering decisions down the road. Given the significance of this decision on the product, as well as engineering aspects it should be debated and decided jointly by the architects and product managers. Choosing battery power is great for ease of installation and flexibility in choosing the installation location for consumers. In return, it limits the available power on the device significantly. A constrained-power design is very different from a liberal-power design. The former also requires the user to replace the battery once in a while, which impacts the user experience. Today in the market, there are ample choices for both types given enough demand for both use cases. Once the architects and designers make the skeleton decisions for the shape and functionality of the electronics, they document, freeze, and send it to the engineering teams for development.

The hardware will include electrical modules for audio and video processing, memory that is sufficient for storing a reasonable amount of video, radio communication modules, sufficient processing power per design specifications, and power supply. The architects must provide the high-level specifications for these modules. On the mechanical side, the design should tolerate sunny and windy situations while being ergonomically designed for the person who pushes the button. Since it is designed for consumers the aesthetics are also of importance.

The radio design needs to handle two scenarios. In steady-state working conditions, the module needs to communicate high-quality video to the network with low latency. With that decision alone we ruled out Bluetooth or Zigbee because they are not well suited for video communication. Given the residential nature of the product, a strong Wi-Fi signal is a reasonable assumption; therefore, we need Wi-Fi modules in the design.

The second radio requirement is for the initial setup of the device; when the device comes out of the box somehow it has to connect to the desired Wi-Fi network. A common technique is to also include Bluetooth radio in the device; once the device is turned on (or otherwise is placed in configuration mode), its Bluetooth device becomes discoverable. By connecting to it using a smartphone, the user can then pass on Wi-Fi credentials to the doorbell. From that point on the doorbell connects to the Wi-Fi network and associates itself with the owner's account. Note that this last step requires authentication and profile infrastructure in the backend system by the manufacturer.

Security and privacy are highly important as always. Communication from the device to the cloud and back to client devices should ideally be all end-to-end encrypted. HTTPS communication is a reasonable choice to provide security for the data in motion. Since users need to have access to their video recordings afterward the service provider should provision cloud storage as well. All such communication needs to happen in encrypted format so that eavesdroppers cannot easily leverage the content. Authentication and authorization services should be taken extremely seriously. Password management is essential; storing passwords in an unencrypted manner is extremely dangerous. Users are often required to choose hard to guess passwords and enable two-factor authentication. Notifying users of changes to their accounts or failed attempts to log in are also other very effective security features.

The cloud infrastructure and user management requires a whole different set of provisions. First of all, the users should be able to register themselves, create accounts, add family members, and manage their accounts on their own. They also need to have access to their videos and security alerts. Such storage and processing infrastructure can grow exponentially over time because of the volume of video files. Typically service providers charge a subscription fee to cover those costs and only keep the videos for a limited time.

The lifecycle and end of life management of the device is the other aspect of the design. Users should be able to easily wipe out all of their personal and private data and metadata should they choose to do so. Another scenario is when new residents move in; the new users should be able to easily configure themselves on the device as if the device is being used for the first time. Over-the-air updates to the firmware and software on the device itself guarantees that the manufacturer has access to its devices in case a security or otherwise improvement upgrade is required.

Another remaining decision to make is how much artificial intelligence and analytical power to include on the device as opposed to moving them to the cloud. In the former scenario, once the image or event is captured, the doorbell itself can make decisions without having to pass on the data all the way to the cloud and wait for the inference. In the latter scenario, major analytics happens in the cloud servers. Each approach has its pros and cons; however, with everything else equal there is always a preference to perform analytics as close to the data (the edge) as possible. With artificial intelligence at the edge, the camera can send notifications to the user faster in case of detecting anomalous movement; in burglary situations, even five seconds can make a significant difference, which can exactly be the difference between edge and cloud analytics. Another benefit is less security and privacy risk; the footage data does not have to travel over the network for the functionality to occur. This can be significant, especially given that there is a level of mistrust of the ways that the tech companies use their users' private data in pockets of their user base. The downsides of this approach are twofold. On the one hand, the design should be more powerful, which adds to the cost of production. It also adds to the cost of security, regulatory, and lifecycle management of the device. The other hidden cost is the extra power that the doorbell needs to burn while doing the inference. Of course, in the case of wall-powered design, the second problem doesn't exist. In the battery-powered systems, the power needs to be compared to the power required to communicate over Wi-Fi.

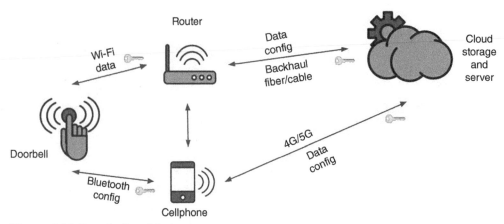

Figure 5.16 Doorbell architecture.

The last part of this example reveals the secondary business opportunities for the product. Once consumers start using the doorbell, they produce lots of data that they may need to keep for an extended period. Offering cloud storage to the customers is a business opportunity. Additional security offers such as fast communication to the authorities when the user sees an imminent threat is another viable service. Figure 5.16 illustrates the architecture and data flow in the product design all the way to the cloud.

Energy-Efficient Office

Many office buildings have been built in an era that energy efficiency and sustainability were not major factors. Modernizing the energy systems of these large commercial buildings is a significant task. Modern energy management systems can perform lots of these operations with a few configurations. The remaining challenge is the instrumentation of the building. For the management systems to operate effectively, they need to at least measure temperature, pressure, occupancy, and submeter electricity consumption. Next, they should be able to actuate the right systems per their optimization algorithm. In this example, we focus on the first part of the problem. We consider a multi-story office building that was built in the 1970s. Besides the simple temperature sensors that are attached to the thermostat in each zone of each floor, there are no other connected temperature sensors. Similarly, submeter electrical measurement is nonexistent. We look at this problem holistically and try to design an IoT system to tackle the problem. We put ourselves in the shoes of a high-level system designer and go through the high-level decision points of a solution.

In the first stage, we divide the problem into two distinct modules. The first part is to gather required data from across the building and bring them in a central place where control happens. The second part of the problem is to generate and apply the right controls.

The control problem in our example is relatively straightforward. Once the temperature and submeter electricity measurements are gathered in the facilities control room, commercial building and energy management systems can then optimize the AC and heating units' operations; while significant, it becomes more of an optimization and control problem. Let's focus on the instrumentation and data gathering part.

The first observation is that an effective energy management system requires temperature measurements from several different spots across the building that are normally not wired. Measuring occupancy is also important, which can be handled through Wi-Fi traffic measurement as a proxy. There are infrared motion sensors in the market that also detect and count the number of warm bodies in their field of view that can be used; the latter is a more complicated and costly solution. Some organizations use a combination of all of the above plus their badge systems to get to an occupancy number as close to reality as possible.

The fact that temperature measurement needs to happen in several locations at the office calls for a wireless architecture. A proof of concept phase can be started first before committing to heavy investment for the whole office. The goal is to get an idea on how much energy can be saved before investing heavily in the whole building. Such a proof of concept phase includes custom wired or wireless sensors for both temperature and submeter electricity measurement. Energy and data analysts can come back after a month or two and make sense of the data. For example, if the data shows that indoor temperature is kept cold on a hot day that is a weekend with no humans in the building they can make a case for a saving opportunity. Another scenario is when the building is kept warm after business hours or too early on a winter day when the occupancy is very low. Older building management systems often work on a set schedule and these scenarios are very common. From there the energy and facilities specialists can make an estimate calculation for energy savings should the operations get optimized. That number will determine the financial viability of the project. A couple of key results of the study phase are how long it takes for the capital expenditure to break even with the energy savings and how much impact on the carbon footprint the company should expect.

After the concept is proven, the design process starts. There are many options for economical wireless temperature sensors already in the market. Bulk prices make them even cheaper. The same goes for submeter electricity sensors that can measure the electricity usage granularly at different times of the day. These sensors typically connect to a central gateway hub through Zigbee. The reason for that choice is that Zigbee allows for low-power mesh networks that enable installation of sensors in fairly arbitrary locations. This radio communication standard provides low data rates, which are more than enough for the temperature measurements of our scenario. It is inherently secure because it allows symmetric cryptography between the two communicating sides. This fact is important when project owners seek approval from the IT department and facilities managers. The gateway hubs then unify and transfer the measurements through Wi-Fi to a central database. That communication also needs to be secured the HTTPS. The database can be part of the service or can be provisioned by the office managers. The latter option may require more skills to develop such a database and maintain it. Compared to our consumer design example, the user-friendliness of the solution has lower priority. It is completely reasonable to assume that system experts with sufficient technical knowledge will install and operate the system.

The human analysts can analyze the data and make optimizing decisions based on all the gathered data. For example, it is common to realize that certain zones in the building have been unnecessarily cold or hot and thereby burning energy while keeping the occupants uncomfortable. These kinds of discoveries lead to high-level changes to the way the building energy system works. Similarly, submeter electricity sensor data identify high-electricity spots in the office that may be a result of employees trying to use personal heaters or run big servers. These discoveries can help remove unnecessary loads or potentially move them to the off-peak hours. It is very common that after data collection becomes possible new revelations are made which will in turn lead to further system optimization.

The real-time operation of the building energy system is left to the automated systems. Once high-level discoveries and opportunities are exhausted by the human analysts, an automated system connects to the cloud or the server hosting the data. From there it can run its own learning algorithms in batch mode and make instantaneous decisions in real time.

The lifecycle management of the system requires facilities and IT staff to keep the gateway and backend database systems up to date. Security flaws are discovered all the time that need to be patched. For larger-scale projects of several hundred sensors or more, an IoT platform can be deployed to monitor the health of all the assets, see their data flow, provision new devices as new needs arise, add context to each data stream for easier analytics, and also monitor unrecognized attempts to connect to the network. These IoT platforms can also serve as the backend database to be integrated with the rest of the IT system.

IoT SYSTEMS LIFECYCLE

To conclude this chapter, we briefly review the important lifecycle considerations for IoT devices; the life starts when a device is placed and provisioned in the network all the way until they need to be decommissioned.

1. **Provisioning** is the process of initial installation and powering of the device. The very first steps of the configuration can also happen here. This phase includes the physical adjustments to the IoT device so that it is placed in a safe spot that is effective for its purpose.

2. **Authentication** is the next step. When the device goes online it needs to be recognized by the IoT platform, IT-managed Wi-Fi, home network, or anyone who has the authority and responsibility to let a new device in the network. This process can be manual in smaller systems or automated in larger-scale scenarios.

3. **Configuration** is the next step. When a user configures its security camera to only send alarms in certain hours of the day, they are configuring their device.

Similarly, industrial and commercial use cases require configuration en masse; that configuration is either scripted or handled by the IoT platform.

4. **Control** is only relevant for devices that offer the feature (they are not just sensors). A control algorithm that lives on the device or remotely will issue the operational command for control purposes. Open loop control can be configured during the configuration phase, while smarter closed loop control requires analytics at the edge or from a remote server.

5. **Monitoring** refers to the constant observation of the health of an IoT device. Basic health measures include checking the heartbeat signal to make sure the device is on and connected. More sophisticated aspects of the health detect when data look anomalous or wrong through statistical measure. Security measures are sometimes applied in this process to detect compromised IoT endpoints.

6. **Diagnostics** is the natural next step after monitoring. When an issue is suspected the diagnostics kick in to pinpoint the type of the issue. Typical issues vary from low battery, hardware malfunction, and network problems to more sophisticated problems arising from cyberattacks and spoofing. While simpler problems can heal themselves using resilient networks and devices, more complicated issues will need support staff.

7. **Software updates** are an absolute must for devices that are even slightly smart. The rate of security attacks is so high that security patches are issued regularly. Over-the-air software updates can also offer an economical way to improve the functionality of the device through improved software.

8. **Maintenance** refers to the totality of the steps that are taken to maintain an IoT device over its lifespan. Besides monitoring, diagnostics, and software updates, physical maintenance of the devices, as well as procedural warranty checks, set-point adjustments, and caring for the devices all make up this category.

9. **Decommissioning** is the final stage and happens when economical or security justifications for keeping a device online disappear. It can be because of the new needs in the organization, unpluggable security holes, old age of the devices, emergence of a new technology, or changes in the business circumstances. It is important to not overlook private or sensitive data that may be retrievable from the memory of many IoT devices that are dumped without proper cleansing. Devices should be carefully wiped clean of any sensitive data before discarding. Sustainability and environmental considerations are other important considerations when discarding a large number of electronic devices.

Chapter 6

The Process of Building a Data Product

So far we covered multiple aspects of an IoT product or service including its business viability, software and hardware design, security, networking, and lifecycle. Almost all IoT projects have a significant data aspect to them. The data can be the essential element of the whole project; for example, when an airport wants to measure the efficiency of the water heaters across all of the terminals, the hardest part may be collecting temperature data. In some other cases, data opens the door to secondary opportunities for add-on business models. For example, when a smart doorbell manufacturer offers paid storage of video footage for a year for its subscribers, they are inventing a new revenue stream on top of the original doorbell functionality.

In any case, data is the blood stream in the body of most IoT systems. For that matter, we dedicate this chapter to the concept of building a data product. Our focus here is not on the hardware, networking, or algorithmic aspects. Rather, we will take a look at the process, personas, and the right ways to tackle the challenges. The specific technical and business decisions will become results of this process. You would be surprised to know what percentage of large and small organizations fail at the basics of this process and, therefore, waste significant amounts of time and money. The key

IoT Product Design and Development: Best Practices for Industrial, Consumer, and Business Applications, First Edition. Ahmad Fattahi.

takeaway is that just having the right intentions of solving business problems with data is not even remotely enough to guarantee success. In what follows, we look into the personas needed to be included in the right team, emphasize the important role that subject-matter experts play in the process, and how the whole process is far from a smooth linear flow.

In going through this process, we leverage an established high-level framework called CRoss Industry Standard Process for Data Mining or CRISP-DM (Figure 6.1). This standard process was originally built in Europe in 1996 for what used to be called data mining projects. Even though the technological landscape has advanced dramatically since then, the core principles of this process still hold strong. You may find several other competing processes that look very appealing, more detailed, or a better fit for your project, and that is absolutely fine. We study CRISP-DM because of its simplicity and universality. We understand that any project will have corner cases that would not exactly fit this process. By the end of this chapter, we will have gained a good understanding of how to approach a real-world data problem, how to form the right teams, what expectations to set, and what stages to anticipate.

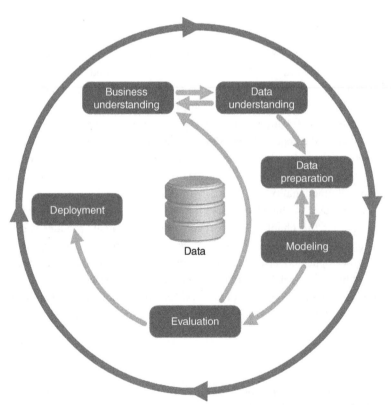

Figure 6.1 CRISP-DM process diagram (*Source:* https://en.wikipedia.org/wiki/Cross-industry_standard_process_for_data_mining).

WHAT PERSONAS DO YOU NEED ON YOUR TEAM?

Having the right skills and personas in the team can easily be the difference between a successful or failing project. For data products and services, the need goes well beyond data scientists and engineers. Let's start going through the CRISP-DM process and list the types of personas per each box. Starting from *business understanding*, we need business developers, subject-matter experts, and product managers who can connect the real-world problem with a technical data problem. This is where a problem such as "How do we increase the yield of this semiconductor fab?" gets translated to "We need to use test data from various stages of wafer and die production to classify and correlate them with each stage."

The next box is *understanding the data*. We need data architects, data strategists, and analytical experts to join forces and make sense of the existing data. These personas go deep into the meaning of each data set and, hence, need to be well aware of database, security, quality, and other aspects of data sets in connection with the problem at hand. The next box is where *data preparation* happens. What's needed here is mainly data engineers who are highly skilled with the engineering aspects of the data. They work closely with data strategists to access and shape the data in the right way.

Modeling is what most observers expect to see in a data science or machine learning project. Business intelligence specialists, statisticians, data scientists, and mathematicians are needed here to put a structure on the data and build models per the technical description of the problem. Model evaluation is sometimes done by the same personas especially in smaller projects. In larger projects, there are dedicated quality and validation teams who validate the produced models.

The next box, labeled as *evaluation*, is when the produced and tested model is evaluated against the initial business intentions behind the project. To the less experienced eye, this step may sound frivolous because everything started from the business problem. However, getting to this point is often a long and meandering road from the start of the project. During this process, things veer off, sometimes to very strange places. So it is crucial to close the loop. This step is completed by a group of the personas from the previous steps including product managers, subject-matter experts, and data scientists. Last but not least, for deployment there is a need for machine learning engineers. This persona is skilled in taking a working data science model and putting it in a production environment. *Deployment* of a model in includes building the necessary data pipelines for the real-time operation of the model. The deployment environment can be the cloud, a few local servers, or an edge device.

To bring all of these pieces together, you need an overall owner who has executive decision power. When conflicts of priorities arise, deadlines are about to be missed, or there is a disagreement in the team, having a final decision maker is absolutely critical. Funding is another major role that the executive owner plays. In order to keep the machinery of the project well-oiled, project management is necessary to keep things aligned and on track.

Depending on the specifics of the project, there may be a need for marketers, launch managers, documentation specialists, lawyers, and supply chain specialists. We are not going into the details of this last group because they are not specific to the CRISP-DM process. Next, we go around the CRISP-DM circle and explain in more detail what each distinct box entails and what pitfalls are common.

BUSINESS UNDERSTANDING

It's all about business. Unless you work in academia or a pure research facility, it is all about business. The process of building a data product or internal project or an external service all starts in a business need. It may sound obvious when you hear it. However, this seemingly trivial point is forgotten time and again when teams get into the weeds. You can hear engineers enthusiastically champion using the newest hardware release. When you keep asking why we should adopt it, sometimes the answer is nothing but excitement about the technology. The same goes for data scientists, data engineers, and typically anyone who is closer to the building process and farther from the business layer. We all have seen numerous situations where a technical person, with a spark in their eyes, proposes using the newest bot to automate a portion of the user experience. It is a great thing to have proactive and enthusiastic team members. However, the initiatives need to clearly align with the business goals.

One of the best ways to get ahead of this problem is to create more overlap between the circle of knowledge in the technical and the product or business teams. That way the technical teams will have at least an idea of the bigger context of the product they are being asked to build. Once that happens, good technical professionals are actually very skilled at judging if a new technology moves the needle significantly or not. For example, imagine that the goal of an internal project is to use machine learning and data to improve the customers' experience with the company's product by a specified measure. In a vacuum, the technical teams can propose tens of ideas with various artificial intelligence algorithms. Are most of them cool ideas? Of course. Do most of them help the customers' experience in a significant way? Usually not.

Now imagine that the technical team spends a minimal amount of time with the product managers and customer research teams. They learn that most of the customers are complaining about the installation process of the product as opposed to how to get support. Once that becomes clear, automatically people pay attention to innovation in that area as opposed to adding a bot to the support process.

Product managers and managers as a whole are in the best position to keep the whole team within the lines. That is yet another reason why a clear definition of the deliverables for any product or service should be published for the whole team to see. Not only that, successful contributions toward product features should be defined in a measurable and quantitative way so that *business value* becomes as objective as possible. At the same time, implementing a process to harness, measure, and adopt innovative ideas from all corners of the team is a great vehicle to have *innovation* and *business discipline* at the same time.

It is important to make sure new ideas and innovation is not frowned upon in the process. Rather, this is a directional recommendation to coax creative minds in the direction of business value. As we will see in the remainder of the CRISP-DM process, this alignment with the business needs comes up explicitly in at least one other pit stop.

DATA UNDERSTANDING

The best thing about being a statistician is that you get to play in everyone's backyard. This is a famous quote by a renowned twentieth-century American statistician named John Tukey. The translation of this quote in the business world today is that a data team gets assigned to solve various problems all the time. While exciting and refreshing for them, it means they have to learn enough about the underlying business problem and data definitions quickly. In Tukey's terms, in order to have fun playing in the new backyard, they need to know where the swing is located and where the pitfalls may be.

This brings us to the extremely important and often overlooked role of the subject-matter experts. The subject-matter experts are the best tools in the arsenal to inform and educate the data engineers and machine learning modelers about the problem they are tackling. Without the right level of subject-matter expertise, data scientists can waste lots of time trying to study the literature, misinterpret the meaning of certain data points, or fundamentally solve the wrong problem. Two practical points are worth mentioning here. First is that the expectation has to be set properly in the beginning by the leaders. Less experienced data professionals may believe, with the best intentions at heart, that they know how to make a model as soon as they have access to the data set. It takes some real-world experience to know how many things can go wrong without the subject-matter experts actively present. The second point is that everyone should expect this process to take some time and energy. Project managers should carve out days or weeks, depending on the complexity of the problem, for this fusion of teams to happen. Managers of both teams, data and business, should also agree that their team members will be spending a significant amount of time together building the shared base of knowledge. It is not uncommon for good data scientists to spend weeks on the factory shop floor observing and talking about the process they are trying to help. Over time, when a data scientist learns the nuances of a certain industry, they can solve problems really quickly. At that stage, those technical resources become extremely valuable resources for any organization.

Note the bidirectional arrow between this stage and the initial business understanding. What it emphasizes is that both of these stages impact each other in a meaningful way. On the one hand, the business vision provides context for the available data. On the other hand, as data discovery progresses, subject-matter experts provide feedback to the product owners. This is the best time to find out if there is sufficient data to deliver a certain feature; if the process goes past this point, several precious weeks or months can be wasted before the technical developers raise the flag about insufficient or low-quality data.

Managers and data scientists alike should also be wary of the optics of their approach. In some situations, the cultural narrative is for artificial intelligence to be responsible for the eradication of many good paying jobs. That mentality can create a wall of mistrust and fear-based resistance between the two sides. The subject-matter expert may see the data team as a preamble for him to be replaced by an algorithm. While there may be some truth about this mentality, the level of threat is often exaggerated. In most use cases, the algorithms come to the assistance of the humans to help them be safer, more successful, or take a repetitive and boring part of their job off their plate. The intentions of the project and the end game should be explicitly communicated to the subject-matter experts in advance. Waiting for them to speak up is often not an effective way of managing this situation. In general, the territorial nature of this phase of the project can be a challenging situation that requires creative solutions and human skills by the leaders.

DATA PREPARATION

Once the project owners and data architects and strategists converge they ask data engineers and database administrators to prepare the data for modeling. The guidelines are written and signed off by-product managers and architects. This is when actual features are engineered and permissions for accessing data at scale are sought. For example, imagine a smart wearable device that is supposed to offer a new feature. The new feature uses personal data from each user's average number of steps during a week. The number for each weekday is then combined with several other features in a model to generate a reminder push notification for the user to move or join a training program at a time that is personalized for each user. While this feature looks quite simple, in reality the *average number of steps per each weekday* may not exist in the database exactly as needed by the model. That is where the data engineers come in to engineer this feature. Many challenges can arise for this seemingly simple feature. What should be the time span of averaging? How should the feature handle missing data? What if there are out of size anomalies that can unilaterally impact the average number of steps? Do various countries impose various regulations when it comes to accessing users' personal data? How often does this feature need to be refreshed and recalculated? What is the most efficient way of recalculating the feature to minimize the burden and cost on the servers? When is the first time that this feature should be calculated for newly joining users? These are all real questions that an engineering team needs to answer before they can actually build the data pipelines.

In many use cases, the end of this stage marks a significant amount of advancement for the data product. The actual modeling and deployment, while very important, are often clearer tasks with less ambiguity than the first few stages. Unless the artificial intelligence requirements are cutting edge, existing open-source packages and commercial tools offer several options to build the model. To be clear, the rest of the way is still absolutely important and nontrivial; however, it is often much less ambiguous. Getting to a point where the real-world problem is clearly agreed upon, it is translated to a technical problem, and the individual features for the model are defined and built is often the more treacherous part of the process. The end of this stage is when everyone starts to *see* what the end game will look like. Be prepared for the teams to spend several weeks in this stage.

A potential pitfall for this stage is when access to the data is a challenge. Up to this point, managers and architects assume access to certain data and build upon that assumption. Data engineers may run into organizational silos in case of internal projects. These silos become increasingly likely the larger and older the company becomes. External projects dealing with personally identifiable data have to deal with a higher level of sensitivity and regulation that needs proper attention. Also, some models need the feature refreshed once in a while, for example once a week. Some other models require almost real-time access to the features as they happen. From the engineering perspective, these two setups impose very different architectures and challenges.

MODELING

Building the model is what most outsiders imagine when thinking about machine learning. This is the part where a data scientist learns about the problem, receives the features from the data engineers, and goes to work. The work is building a mathematical representation of the data. The question to be answered can be of various types. One class of models, called supervised learning, requires labeled data so that the model can learn from it. The process of labeling the data can be a significant hurdle or cost for supervised algorithms especially in case data volume is large. There are certain online platforms that connect data owners with volunteers or paid contractors to label data masses at low prices. Besides the cost, you may need to be wary of the quality of labeling when using such services.

The most common type of model is a predictive model. In these supervised models, the goal is to predict what will happen at a future time based on historical observations. An example of this is predicting if a piece of equipment will fail over the next 24 hours given the noise it's making today as well as its age, maintenance history, and workload up to now. Depending on the nature of the prediction problem and the algorithm, the required data volume varies significantly. You typically need at least a few hundred observations for the simplest models to do a decent job of prediction. More sophisticated and larger models, such as neural networks, need many more data points.

Another class of models cluster observations into a finite number of similar groups. Similarity can be a subjective concept. Therefore, the modeler may try several different approaches and compare the results. Some algorithms need the data scientist to specify the number of clusters as an input which is clearly a human-in-the-loop approach. An example of this type of model is when hundreds of wind turbines share their operational data over time with a server. The operators are interested in knowing how many classes of operation exist in their wind farm. For example, a quarter may be operating at close to 50% of their nominal output, another quarter close to 70%, and half about 90%. Clustering problems can be multidimensional taking into account multiple variables. They can also reveal anomalous points that only belong to a cluster of a few remote points.

The next class is classification models. These models are often supervised, meaning that humans should label a sufficient number of observations for the model to get trained and start classifying observations properly. These models have many applications when it comes to image recognition. For example, an IoT device with a camera can be configured to disarm the security system when it detects one of the familiar faces with which it is configured. The model is constantly looking for faces in run time. As a result, it classifies the incoming image either as *match* or *not a match*. These types of models have many applications in time series and trend analysis for process control as well.

There are several other types of models in the literature. Natural language processing (NLP) models extract insights from audio or text data. Smart speakers and all voice-enabled IoT devices are equipped with NLP technology. Causal models try to connect the outcomes to the actual causes based on data. Causes can get easily confused with predictors which are usually built based on correlation. It is important to not assume any good predictor is necessarily a cause for the predicted feature. This brings us to an important concept called model explainability.

While in most use cases the first priority is to be accurate when performing a prediction or classification, it is also crucial to explain why and how the prediction is made. Imagine that a mobile health monitor predicts an imminent heart issue for the wearer. Causing such a high level of distress on the user should be warranted only based on solid reasoning. In such medical applications or where regulation demands transparency, the model should be capable of showing its logic. Not all models offer the same level of explainability. There are some indirect techniques to represent causality with things such as feature importance in prediction. While insightful such measures should be taken with caution before getting implemented as the source of causation. Loosely speaking, the more complicated a model gets the more accurate it gets and the less explainable it becomes. For example, a linear regression model can represent a simple linear relationship among a handful of parameters. It can also explain the relationship through the readable linear equation and the resulting coefficients. However, as soon as the relationship among the features becomes nonlinear and the number of features grows larger, more sophisticated models may be required to model the data. Neural networks offer a well-studied approach to perform well. However, they are very difficult to interpret and explain.

Back to our process discussion, this stage of CRISP-DM is handled by the data scientists and analysts for the most part. The role of the subject-matter experts is also pronounced here because they need to interpret features and guide the data scientists in the right direction. Again in this stage, the process can go awry if subject-matter expertise lacks. The data engineers work very closely with the data scientists to explain the ways features have been built in detail. They are also needed to go back and adjust or build new features as the data scientists discover new needs.

Once the model is built on training data, it has to be validated on a portion of the data set that has not been used for training. The reason is that a model with enough degrees of freedom can learn any data set arbitrarily well. By trying the model on a not-before-seen data set, we make sure that the model has learned just enough to represent the behavior of the system as opposed to every single jitter of the noise in it. We covered this point in more detail earlier in the book when we discussed overfitting in the design process.

EVALUATION

When the internal validation of the model is completed satisfactorily by the data scientists, they push it to the next stage for business validation. This is when the data professionals say "we think we have done what we were asked to build." Ideally nothing in this phase should be a surprise to the line of business because the data and modeling process has been done in close collaboration with the subject-matter experts. However, it is absolutely necessary for the resulting model to be vetted by the product and business owners and for the their sign off to be sought. The main challenge leading to this necessity is that things get lost in translation and throughout the process. The time when the product owners convey their desires to the data architects and strategists can be months away from the moment that the model is ready to be validated. Similar to the game of telephone, the result can potentially be miles away from the original intent. All of the feedback loops and bidirectional arrows in the process are designed to reduce the chances of drastic deviations. Nevertheless, an official presentation and validation is required before models are deployed.

An example of this phase is when data scientists construct a model that decides when crops need to be irrigated. It takes data from multiple moisture sensors, weather forecasts, and user settings about the crops and generates a decision about irrigation. The product owners have a number of quality criteria that need to be validated. The first question is how well the model makes its decisions. The recommended approach for the validators is to have a set of scripted test scenarios and accompanying data sets to test the model. Another set of validation points concern the shape of the model. The types of input features that the data scientists assumed available at the time of modeling may or may not be feasible in a real-world scenario. For example, some geographies may notoriously defy weather forecasts as opposed to the 10-day reliable forecast that the data scientists have assumed. Another aspect is the explainability requirements. Water and the crops are both precious commodities; therefore, product owners may want to specify a degree of explanation to the user as to why "water now" or "don't water now" was decided by the model. It is quite possible that product owners decide to sacrifice a certain amount of prediction accuracy in the interest of more explainability.

As you see in the CRISP-DM diagram, there is an arrow moving from this stage back to the business understanding. While data professionals work on the clearly defined technical problem, they may uncover new behavior, unforeseen challenges, regulatory issues accessing data, or otherwise find totally new opportunities in the data set. As a result, the outcome can inform product owners of a renewed challenge or opportunity depending on the finding. Therefore, a new cycle can start to address the challenges or harness the new opportunity. A real example of this scenario was when a large mining corporation equipped their massive hauling trucks with connected instruments. The initial goal was to measure the shock on the wheels, keep an eye on the fuel and oil levels, and other similar mechanical aspects. What they surprisingly discovered was some major behavioral inefficiencies by the drivers. They discovered that more than 10% of the drivers drive with parking brakes engaged. Many other ones did not turn off the engine while at long stops against the company policy. As a result of these findings in the validation phase, they kick started new training and process improvements to correct the behavioral mistakes.

DEPLOYMENT

Once everything checks out, the model is ready to be deployed. Deployment means that the resulting machine learning asset moves from the development environment to the production system. This is a significant task that requires skilled machine learning professionals and data engineers to work closely with the data scientists. Below are some of the main differences between a development and production environment that needs to be addressed.

The development is optimized for the training of the model per the requirements. Inference during the runtime is a significantly different task that requires new ways of optimization. For example, if a model is supposed to make 10 million inferences a day, it is very important that it is agile and small enough that a large number of inferences doesn't incur massive cost. The implication of this fact is that lazy models that leave major parts of inference to the decision time are ruled out. On the flip side, if the volume is really low, optimizing the models to be efficient at runtime will not be a high priority.

The models that run on real-time data require data pipelines that provide the data in real time for inference. In the development phase, data scientists can make assumptions or feed in simulated data points. However, a production environment should provide streams that meet all security and latency requirements.

When a model is ready and trained, it has to act on the real data that may come from various parts of the world. Different countries have varying levels of data privacy regulations. Some of those regulations prohibit personally identifiable data to leave political or jurisdictional borders. That limitation may require the resulting model to be deployed in multiple locations so that they can act on local data.

In the case of edge computing and small devices, the resources are limited. While a model can be perfectly fine in the development environment they may run into resource issues on the small form factor in presence of other pieces of software. For example, a smart home company built smart algorithms for energy management in the buildings so that they can run on a Raspberry Pi. That way, they could keep the implementation cost low. While the memory requirements were met they realized that the circuit boards get increasingly hot due to the heavy processing at runtime.

The takeaway is that this part of the process is very critical and should not be taken lightly at all. The last mile can still make or break a project.

THE CYCLE REPEATS – CRISP-DM

We went through the whole cycle of CRISP-DM once. We saw several examples of how each stage is important, which personas should be involved, and what can go wrong. After all of these steps are finished, the whole cycle keeps repeating as long as the product or service has business justification. When a network of sensors provide their readings about the pressure of gas in a network of gas pipes, the model keeps inferring if a leak exists or not. As time goes by, real and new data is collected that may be different in nature from the original data that was used to train the original model. That opens an opportunity for the whole team to optimize the model with the more relevant data.

Besides that, the physical environments are constantly changing. Pipes, pumps, sensors, and the substances inside the pipes age or otherwise change. Sensor malfunction happens all the time. That means that monitoring of the model behavior and its quality should never be neglected.

On the opportunity side, when new data and accompanying analyses are offered to subject-matter experts over time, they may make new discoveries or come up with new ideas. Those deeper ideas can unleash a brand new project to create new features. Macroeconomic, technological, and business factors also impact the direction of products all the time. When a new algorithm becomes open source that can perform the same level of prediction at a fraction of the model size, it might warrant a brand new project. If a new partnership is signed with a cloud provider, the business may decide to move all of their models and deploy them in the new cloud environment. New hardware and sensors coming to the market can also pose new opportunities. As long as there is business justification for a product or feature the full cycle keeps repeating itself around the CRISP-DM circle.

Chapter 7

Concluding Remarks

Connecting machines opens the doors to many possibilities. The old adage of a team becoming larger than the sum of its individual members absolutely applies here. Individual sensors, servers, controllers, and servers can do valuable things in isolation. However, when they are connected with each other, through data and control, the value compounds. Almost every report and analyst publication points to massive opportunities in the consumer and business markets for the Internet of Things (IoT) for several years to come. If done right, the IoT projects can unleash trillions of dollars of business value over years, help with the environmental challenges, improve safety and reliability of systems, and make human life better overall.

IoT encompasses a large array of technologies and offers a horizontal capability across several verticals. The budding use cases include smart homes, agriculture, smart manufacturing, power and utilities, transportation, smart cities, and personal wellness. The parallel advancements of hardware technologies, analytical algorithms, public attention to resource efficiency, and connectivity fabric across the globe make the ground as fertile as ever for IoT to deliver value. There is a long and wide value chain with many opportunities to harness. From the niche sensor market to networking, wireless providers, hardware and chip makers, security professionals, and data scientists and analytics, the ecosystem keeps growing in depth and breadth. There is room for vertical players who offer specific end-to-end solutions for niche use cases. There are also good opportunities for horizontal players such as satellite communication providers to serve agnostic needs across all industries.

IoT Product Design and Development: Best Practices for Industrial, Consumer, and Business Applications, First Edition. Ahmad Fattahi.
© 2023 John Wiley & Sons, Inc. Published 2023 by John Wiley & Sons, Inc.

As large as the opportunities are, so are the challenges. Making all of these disparate systems in the ecosystem to connect and work with each other is no easy feat. The whole effort of Matter, which tries to bring interoperability to the smart home ecosystem, is a good example of that. Making standards built in ways that are broad enough to cover sufficient amounts of business interest takes a long time. On top of these challenges, there are regulatory considerations that are equally important. Given the nature of IoT projects, data can travel across political boundaries. Governments and public sector regulators are, for good reasons, sensitive to data sovereignty for national security reasons. The increasing rate of security intrusions over the past years exacerbated the problem.

We anticipate that the tide will get stronger, the challenges will be handled one way or another, and the forces of economic growth will push IoT forward strongly. Regulators will understand the value to the whole society and will get behind them. Ethics advocates will also continue to push for the public interest from their angle to encode users' rights into laws. Similar to most other massive technological shifts, it will be messy but the progress is inevitable.

Further Reading

Annaswamy, S. (2020). Soft PLCs: The Industrial Innovator's Dilemma. https://iot-analytics.com/soft-plc-industrial-innovators-dilemma/.

Chase, J. (2019). Do You Need a Smart-Home Hub? *Wirecutter.* 17 May 2019. https://www.nytimes.com/wirecutter/blog/do-you-need-a-smart-home-hub/.

Higginbotham, S. (2020). Everything You Want to Know about Project CHIP. 14 September 2020. https://staceyoniot.com/everything-you-want-to-know-about-project-chip/.

Wikipedia Diffie–Hellman Key Exchange. https://en.wikipedia.org/wiki/Diffie%E2%80%93Hellman_key_exchange.

Project-chip/connectedhomeip. Github. (2022). https://github.com/project-chip/connectedhomeip.

Frequently Asked Questions about Privacy, Nest. (2022). https://nest.com/privacy-faq/.

IoT Product Design and Development: Best Practices for Industrial, Consumer, and Business Applications, First Edition. Ahmad Fattahi.
© 2023 John Wiley & Sons, Inc. Published 2023 by John Wiley & Sons, Inc.

Index

*IoT Product Design and Development: Best Practices for Industrial, Consumer, and Business
Applications*, First Edition. Ahmad Fattahi.
© 2023 John Wiley & Sons, Inc. Published 2023 by John Wiley & Sons, Inc.

VPN, 82–83
vulnerability disclosures, 92

W

Wi-Fi, 148–149
WiMAX, 153–154
wireless communication protocols, 144–155
   long-range wireless protocols, 151–155
   short-range wireless protocols, 145–151

WirelessHART, 150
work-life balance, 98
wrapping, 71

Z

Zigbee, 147
Z-Wave, 148